Agrobiotechnology in Perspectives

History, Economy, Science and Technology on the Plate

Agrobiotechnology in Perspectives

History, Economy, Science and Technology on the Plate

Edited by

Hon-Ming Lam

The Chinese University of Hong Kong

World Scientific

NEW JERSEY • LONDON • SINGAPORE • BEIJING • SHANGHAI • HONG KONG • TAIPEI • CHENNAI • TOKYO

Published by

World Scientific Publishing Co. Pte. Ltd.
5 Toh Tuck Link, Singapore 596224
USA office: 27 Warren Street, Suite 401-402, Hackensack, NJ 07601
UK office: 57 Shelton Street, Covent Garden, London WC2H 9HE

Library of Congress Control Number: 2023951269

British Library Cataloguing-in-Publication Data
A catalogue record for this book is available from the British Library.

AGROBIOTECHNOLOGY IN PERSPECTIVES
History, Economy, Science and Technology on the Plate

ISBN 978-981-12-7672-9 (hardcover)
ISBN 978-981-12-7673-6 (ebook for institutions)
ISBN 978-981-12-7674-3 (ebook for individuals)

For any available supplementary material, please visit
https://www.worldscientific.com/worldscibooks/10.1142/13423#t=suppl

Typeset by Stallion Press
Email: enquiries@stallionpress.com

https://doi.org/10.1142/9789811276736_fmatter

Preface

"You are what you eat", but how much do we understand the food crop we eat? As a scientist, I research crops down to the molecular level in the laboratory. Thinking out of the test tube, behind the food on the plate, there are stories on history, economy, science, and technology.

Flavour is one of the first impressions we have of our food. While tasters care about the flavour, nutritionists emphasize the nutritional value and the impact on human health. The value of food crops also refers to the economic value, with seed trading being a big economic issue worldwide. To come on the market, the production and quality assurance of the crop products are issues of concern. To ensure productivity, agrobiotechnology is employed. The production of food crops directly influences the environment, and it is inevitably influenced by the environment. Moreover, food choice can be a cultural issue with its historical background.

In this book, food crop is narrated from various perspectives including food industry, nutritional science, marketing, economy, agriculture, environmental science, as well as culture and history. We study a seed, and it shows us the world.

Acknowledgement: Dr. Yee-Shan Ku assisted in preparing this preface.

Hon-Ming Lam, Ph.D.
Choh-Ming Li Professor of Life Sciences
The Chinese University of Hong Kong
April 2024

https://doi.org/10.1142/9789811276736_fmatter

Contents

https://doi.org/10.1142/9789811276736_0001

Chapter 1

From Crops to Flavours

Shwu-Pyng Joanna Chen

State Key Laboratory of Agrobiotechnology,
The Chinese University of Hong Kong
joannaspt.chen@icloud.com

Global food trend has been moving towards reducing meat consumption and replacing meals with plant-based diets in recent years for several reasons. First is the environmental reason. Green House Gas emission (GHGe) has caused global climate change that threatens food security. Animal agriculture particularly cattle farming has been identified as the largest contributor to the GHGe. Second is the sustainability of food supply. Global population growth has outpaced agriculture supply. Farmland reduction due to urbanisation, natural disasters (fire, drought, flood) and biodiversity change is clear and inevitable. Third is the consumer concern on health and wellness such as the issues on blood cholesterol level, weight, blood pressure, sugar and salt reduction, aging (antioxidant effect), and gut health (dietary fiber). Other possible reasons are concerns about animal welfare and religion.

1. Animal-free, Plant-based Foods

Oxford climate change group conducted research in UK (Scarborough *et al.* 2014) and quantified dietary GHGe of meat, fish, vegetarian and vegan eaters in descending order of 7.19 kg, 3.91 kg and 2.89–3.81 kg CO_2 emission per day based on a standardised 2000 Kcal daily diet. Reducing meat consumption and replacing it with plant-based foods has become environmentally driven practices. Globally, different companies launched different animal-free products offering alternatives.

Example 1 Animal-free dairy products

The animal-free dairy products offered by US California based Startup Perfect Day in 2014 — *cow milk without the cow*. It uses bacteria, fungi DNA, and gene pathway with fermentation to produce milk proteins. The product can be claimed vegan with real taste, texture, nutrition of dairy and no lactose, cholesterol, hormones or antibiotics concern. The proteins are claimed to be identical to that found in cow milk.

Example 2 Animal-free egg products

The animal-free egg products. Such products already exist for sometime. Many companies such as US based Just Egg, Zero Egg, Beyond Eggs, and others of different country of origin like India, France, Poland, and Singapore produce them in a liquid ready-to-cook format. A blend of bean protein and other ingredient are used to make these products tasty. These products can be claimed vegan, no cholesterol, have no food safety issue (salmonella contamination) and have long shelf life. For example, the ingredient list of Just Egg Vegan is water, mung bean protein isolate, canola oil, dehydrated onion, gellan gum, natural carrot extract, natural flavours, natural turmeric extract, potassium citrate, salt, soy lecithin, sugar, tapioca syrup.

Example 3 Plant-based meat products

The plant-based meat produced by Impossible Foods (burger meat without animal meat). The beef patty is made from blends of soy/potato protein and other ingredients. To make the meat patty taste meatier and closely resemble real beef bloody sensation, soy leghemoglobin is added. This is the heme protein existing in soybean nodules for nitrogen fixation. The ingredient list of Impossible Burger: water, soy protein concentrate, coconut oil, sunflower oil, natural flavours, <2% of potato protein, methylcellulose, yeast extract, cultured dextrose, modified food starch, soy leghemoglobin, salt, mixed tocopherols, soy protein isolate, vitamins, and minerals.

Finally, the milk alternatives using different plant protein as drink basis are available commercially and are growing in popularity. These include soy milk, almond milk, oat milk, coconut milk, peanut, and others. Original or flavoured, simple protein or protein blends exist for direct consumptions. Barista product range was also created for coffee and tea applications. All these plant-based products aim to reduce animal carbon footprint and bring healthier choices to consumers.

2. Crops for Plant-based Foods

Many crops can be used to make plant-based foods and drinks. The commonly used ones are grains (wheat, rice, maize, oat, barley, millet, sorghum, chia, flaxseed, ancient grains like quinoa and others), legumes (soybean), pulses (assorted beans like black, mung, fava, pinto, navy, kidney beans, lentils, and peas), nuts and seeds (almond, peanut, cashew, walnut, hazelnut, rapeseed, hemp), fruits and vegetables.

Crops are versatile. They can be consumed raw as-is or cooked. They can be further processed (pressed, extracted, fractionated) to make oils, juices, and flours. Many protein-containing crops can be made into protein fractions such as protein concentrates, protein isolates, texturised protein, protein meals, peptides, and further into dietary fibers, sweeteners, stabilisers, emulsifiers, thickeners, and other by-products. These fractions become common industrial ingredients for food and feed formulations to bring nutrition, mouthfeel, taste, and body (a sensory term) of the products.

Soybean crop as an example can be used in many ways. It is the best example of a versatile crop. Cooked green soybean like edamame is a famous food. Soybean is used to make soups and stir-fry blends (as beans or via growing bean sprout). It can be roasted to make a snack. Soybean can be processed to make fresh or shelf-stable soymilk, tofu bean curd, various tofu-skin foods, and soy flour for further fractionation. Fermented soy products are regularly consumed such as natto, miso, soy sauce, tempeh, Chinese FuRu, Japanese and Korean bean (jiang) sauces. Fractionated soy products include soybean oil, soybean meals, protein concentrate/isolate, peptides, dietary fibers, and lecithin. They are the most widely used protein ingredients in making plant-based foods and feeds. Even a production by-product, Okara, after making soymilk and tofu can be used as foods, food ingredients, and feeds after drying.

Crop's utilisation and commercialisation brings expectations. **Industrial users** expect the crop and crop ingredients:

1. to have sustainable supply (quantity, availability, and affordability),
2. to provide consistent quality (variety, flavour and taste, stability, and performance),
3. to be safe (regulation complied, contaminants avoidance on mycotoxins, pesticides, heavy metals), and
4. to be environmentally sound (no deforestation, ethically farmed, good land management, caused least pollution, good biodiversity, and have traceability).

Consumers, while seeking plant-based diets anticipate these foods:

1. to be allergen free (allergens include milk/soy protein, gluten, peanut, and others),
2. to have good flavour, taste, texture and colour,
3. to be highly nutritious (protein, sugar/fat balance, nutrients),
4. to be Non-Genetically Modified Organisms (NonGMO),
5. to carry clean labels (contain limited ingredients that are well known), and
6. to be natural and preferably organic.

Sustainability becomes a key expectation in recent years for environmental protection. All these expectations and/or requirements are making the plant-based food supply chain complex.

3. Genetic Modifications in Crop Improvements

To cope with global fast-growing population and increasing food security issues, agricultural output must improve to produce high-quality and high yield food crops. GM (Genetic Modification) technology has played a significant role in increasing the quality and quantity of soybean production. Since 1996 Monsanto Roundup Ready^{TM} (herbicide resistant soybean variety) introduction, GM soybean varieties with different traits were developed over the years. Nowadays, three generations of GM soybeans exist including traits in herbicide tolerance (HT), insect resistance (IR), environment endurance, nutrition, and quality improvements. The first generation offers single trait (HT), double stacked (HT and IR) and multiple stacked (herbicide, insect, weed and pest). The second generation offers soybeans that are tolerant to drought, salt and cold. The third generation offers nutrient enhancement such as higher oleic acid and omega-3-fatty acid (ISAAA, 2016). A Chinese University Hong Kong team under Dr. Hong Ming Lam decoded a salt tolerance gene in a wild soybean (Qi *et al.*, 2014) and subsequently bred three stress tolerant soybean cultivars that can grow in China's Gansu Province, a water stressed area. This enabled soybean production to provide income for smallhold farmers and to improve soil quality in the area.

Tomato variety development using genomics approach is another good example for food crop improvements. Studying hundred tomato varieties uncovered over 200,000 genome structural variations. By applying gene editing techniques, more tomato variety variations including flavour, fruit size and productivity improvement can be derived (Alonge, *et al.*, 2020). Additional traits related to disease protection, insect resistance, stress tolerance, delayed ripening and nutrition quality improvement have also been developed. All of these contribute greatly to crop production quality and quantity improvements.

Opportunities for future crop development either for researchers or industries are many and available. Using genomics, breeding, biotechnology techniques, more crop variety can be created to increase diversity and production. The new varieties can deliver higher yield with better environment coping abilities. Crops can be developed to grow in different seasons, to yield better flavour and nutrition (protein, fat and others), to offer more functional secondary metabolites as nutraceuticals. By-products of crop manufacturing can be utilised in different ways such as food consumption (soybean Okara) and extraction of functional ingredients for industrial use (lecithin and dietary fiber).

4. Cooking and Chemistry

Majority of crops require cooking to make them consumable and palatable. Cooking is not just an art. It is essentially a chemical reaction process. All food is made up of chemical molecules. Cooking is done by applying heat to generate a series of chemical reaction or transformation yielding smell and taste compounds that form aroma and flavour. A celebrity UK Chef, Heston Blumenthal (Owner of UK Fat Duck Restaurant), called it "Kitchen Chemistry". In his exploration with flavour company Firmenich to understand the aroma compounds generated from cooking, he discovered the magic of kitchen chemistry and made a series of films and books to promote the chemistry which happens in the kitchen among different ingredients (Kitchen Chemistry — Discovery Series 2002 and 175 faces of Chemistry by Royal Society of Chemistry). Utilising chemistry principle, he was able to create multi-sensory dishes to generate a unique eating sensation and make good food pairings. Such achievements led him to win the best Chef in the world award in 2005.

Chemical reactions in cooking involve "among components" reactions such as protein, sugar, and fat interactions (Maillard reaction) and "within components" interactions (sugar caramelisation and lipid auto-oxidation). The most famous Maillard reaction, also known as browning reaction, is the reaction between reducing sugars and amino acids. Under proper cooking condition with the right temperature, time and acidity, the reactions happen and generate amazing mouthwatering aroma compounds such as pyrazines, aldehydes, carbonyls, furans, pyrroles (like what is emitted from roasted beef and grilled bacon). All reactions combined generate mixture of molecules delivering very attractive mouthwatering aroma and taste of cooked foods.

Volatile molecules of different foods differ due to their composition variations. If foods have significant number of volatile molecules in common, they work and taste better together. This is the principle of food pairing as mentioned in the above paragraph. By coupling foods with similar aroma molecules, the best eating experience can be possibly created. Examples are pairing banana with caramel, orange with chocolate but not coffee, melon with Parma ham but not acidic fruits.

5. Flavour — Smell and Taste

Flavour is a combination of smell and taste sensation involving two channels, the tongue, and the nose. Airborne odors go through receptor cells in the nostril activating nerve fibers that connect to the brain (olfactory). Food components enter mouth via taste bud pores/receptors which have nerve endings connecting to the brain (gustatory) (BrainFacts/SfN, 2012). Smell involves volatile aromatic compounds. Taste comes from non-volatile ingredients. Sensation of bitterness, sourness, saltiness, sweetness, and umami felt on different part of tongue relates to taste. Only combing smell and taste, flavour can be completely experienced. The experience of a complete flavour can also be affected by other senses like touch (smooth, silky), sight (appearance, colour), and hearing (crunch, crack) in addition to smell and taste (Mouritsen, 2015).

6. Flavour/Aroma Components: Natural, Chemical, Artificial?

Flavour/aroma is a combination of volatile chemical compounds which exist naturally. There are more than 7,000 such compounds identified today. These molecules can be classified according to their functional groups into alcohols, aldehydes, acetals, acids, esters, ethers, heterocyclic compounds, hydrocarbons, ketones, lactones, oxides, phenols, and compounds containing nitrogen and sulphur such as amines, sulphides, pyrazines (Reineccius, 2006a.)

Each food/crop has its own flavour character. Such flavour character is a combination of different chemical classes in different proportions. Limited number of foods can be represented by only one single volatile compound. Banana, may be typified by isoamyl acetate and grape methyl anthranilate. Such single representation smells far from a complete natural food sensation. Food system made up of different ingredients is even more complex particularly after heating. Take fresh-baked bread as an example. The ingredients, fermentation, and baking process all contribute to generate baked-bread's aroma. A mixture of compounds such as maltol, 2-acetyltetrahydropyridine, (E)-2-nonenal, methional, (E,Z)-2,6-nonadienal, and others contributes to the aroma of fresh-baked bread. No one single compound conjures up the smell. Although 2-acetyl-1-pyrroline has a cracker-like aroma and can be considered a key compound in wheat bread crust but it is far from representing the whole baked-bread aroma sensation (Cho and Peterson, 2010).

7. Flavour Study and Creation

Flavour study and creation is done by trapping volatiles from source materials, separating and identifying them, evaluating compound contributions, and recombining compound selections to create the desired flavouring.

Food and source materials are prepared under specified conditions and the volatiles are collected by extraction methods such as the conventional dynamic headspace method, supercritical fluid extraction or SPME (Solid Phase Micro Extraction) (Reineccius, 2006b.) Collected volatiles are separated and identified by GC/MS (Gas Chromatography-Mass Spectrometry). These volatiles are carried by mobile gas through a column and separated by their affinity to the stationary phase coated inside the column. The eluted compounds appear at different time (retention time) enabling mass spectrometer to detect and identify them individually. Based on the amount present, compounds appear at different time with various peak sizes making quantification possible. A sniffing port (GC-olfactometer) can be installed between GC and MS for smelling and tasting. The purpose is to further understand compound sensory characteristic.

Sensory evaluation is a critical part of flavour development. Olfactory device (olfactometer) installed helps with first hand sensory assessment of the eluted compound. Figure 1 shows a typical GC/MS chart with identified compound peaks, the retention time and peak sizes. Figure 2 is the sensory assessment of the eluted GC peaks using olfactory device approach. Each peak can be smelled by human, eluted for later use, tasted, and described to decide its significance (Heath, 1978).

Creation of a flavouring starts with a natural existing target such as an apple or a strawberry flavour. After learning the volatiles involved and their contributions in nature, important compounds are selected for recombining and reconstructing the target flavour. This flavour thus

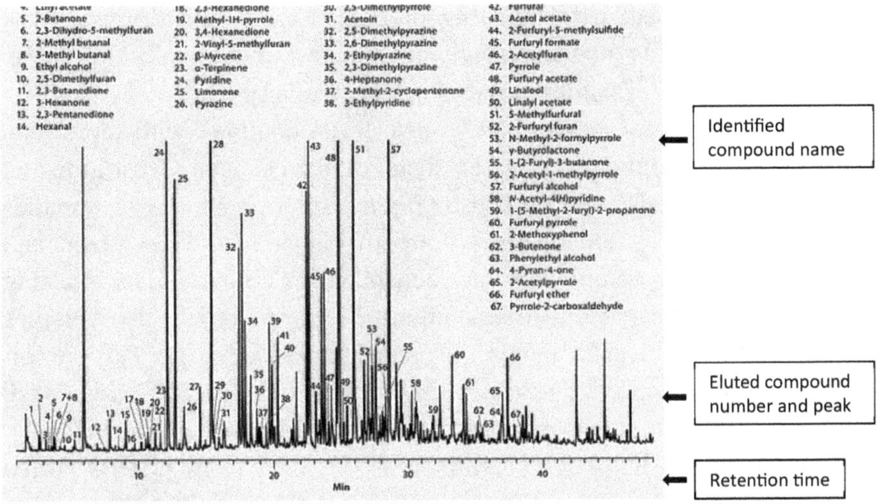

Figure 1. Typical GC/MS chart with identified compounds and peaks.

Source: Accessed on October 2, 2020 from https://www.sigmaaldrich.com/HK/zh/technical-documents/protocol/analytical-chemistry/solid-phase-microextraction/gc-analysis-of-volatiles-in-roasted-coffee-beans-on-omegawax.

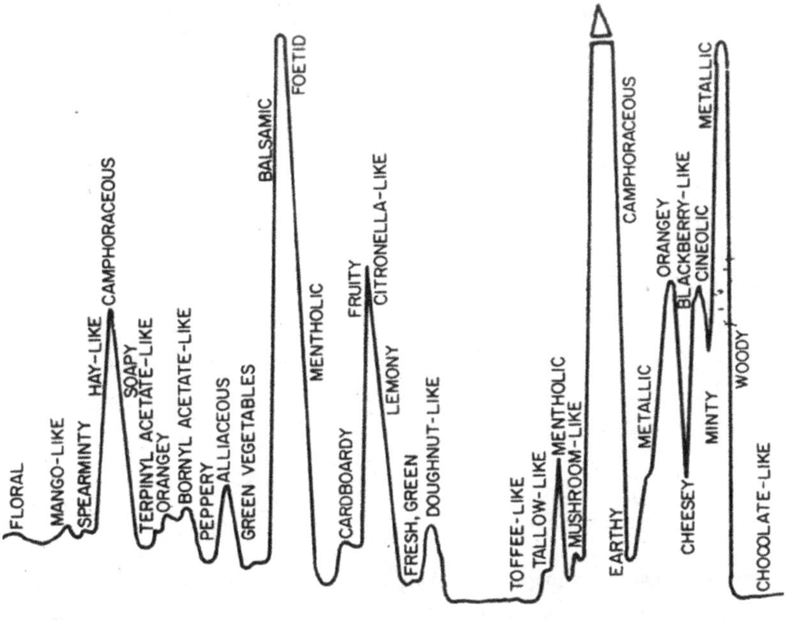

Figure 2. Example of sensory assessment of the GLC eluted compounds.

Source: Heath, 1978, Description of Aromatic Profiles — the culinary herbs, Spanish rosemary oil, Book: Flavor Technology: Profiles, Products, Applications, Section 1.2, AVI publ.co., USA, p21.

created will be used in formulating processed foods and drinks. These selected compounds nowadays are chemicals commercially available, either produced from natural sources or synthetically made.

Coffee aroma compounds are known to be complex with more than 1000 identified chemicals. At green stage, coffee beans are least aromatic with less compounds (~300). Roasting generates many more compounds (>850). Only about 200 chemicals are overlapped between green and roasted coffee. The compounds vary dependent on coffee variety, roasting techniques, and analytical methods. Figure 3 compares the coffee volatiles by chemical function classification at green and roasted stage. This compilation was the cumulative research result from 1958 to 2000, reviewed, analysed and the chart was prepared by Dr. Ivon Flament (Flament, 2002). Viewing the large number of compounds obtained from roasted coffee beans, one learns that reconstructing an authentic coffee flavour is a complex process and task. It is not possible to collect all compounds identified. Not every eluted compound is identified. The compound identified differs when methods of extraction, elution and detection differs. Even by combining all compounds identified by GC-MS will not result in the same coffee aroma because of many reasons (relative concentration and intensities,

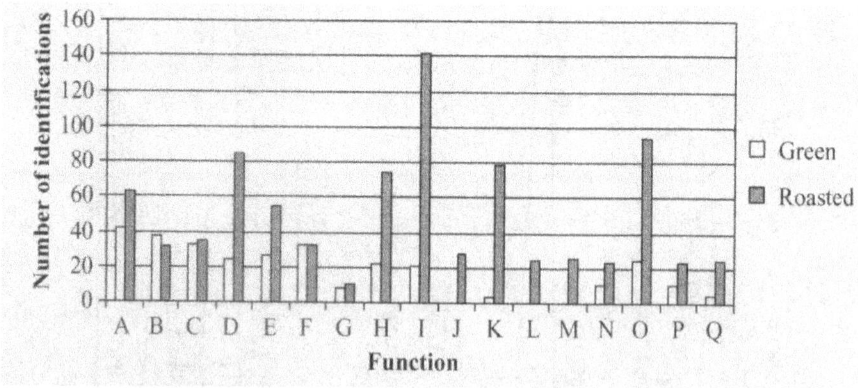

Function: **A** Hydrocarbons; **B** Alcohols; **C** Aldehydes; **D** Ketones; **E** Acids and anhydrides; **F** Esters; **G** Lactones; **H** Phenols; **I** Furans and pyrans; **J** Thiophenes; **K** Pyrroles; **L** Oxazoles; **M** Thiazoles; **N** Pyridines; **O** Pyrazines; **P** Misc. N-containing compounds; **Q** Misc. S-containing compounds

Figure 3. Volatile comparison by chemical function in green and roasted coffee.

Source: Flament, 2002, Coffee Flavor Chemistry, Chapter 4. A historical survey of coffee aroma research. John Wiley & Sons, Ltd., UK., p76.

(extraction or elution methods, reaction and interaction at different stages, sensory variation, etc) . How many compounds and which ones should be selected and combined to create a desired unique authentic coffee flavour is subject to interpretation of the creators and reaction chemists.

Flavouring creation can be done by flavourists or flavour chemists. The former have limited or no chemistry training. Compounds' smell and taste, as well as their interactions with others (molecules or food ingredients) are memorised for each usage. The latter has strong organic chemistry training and may be able to predict chemical reactions and outcomes among different components. Regardless of the level of chemistry training, these creators must have acute sensitivity to tell minute difference of the compounds and combinations. It takes a minimum of five to seven years of day to day training (apprenticeship) with a master to be certified as one. They also need to have good imagination and creativity to generate exciting flavourings which perform in food systems. Making a good flavouring can be a process of repeated trial and error by smelling, tasting and applying.

Although modern technology invented electronic tongue and nose, these can only do partial jobs. Final sensation touch still requires flavourists/flavour chemists to complete. Today, with large data process ability, flavour companies are using AI (Artificial Intelligence) to create new flavours. AI can offer unlimited creativity on chemical combinations, yielding results beyond human imagination. More innovations are made possible and faster.

8. Flavour Language

Sensory evaluation is a critical part of flavour development. All flavours created are required to be evaluated by human panels in food systems via taste and smell to validate their quality. Taste and smell can be very personal experiences. Everyone describes sensation differently because of different sensory threshold, food experience, and language use. A good example is describing a strawberry flavour. When the description is ripe and sweet, to what extent the ripeness and sweetness should be represented in a flavour creation? It is important that a consensus be reached with a set of standardised language to define the descriptors.

Chemical compound dilutions are established as standards to gauge the language used.

Flavour companies build her own set of languages to efficiently communicate with customers to minimise trial and error in flavour offerings. Firmenich Aromasphere® is a global flavour language set covering 155^+ descriptors and 700^+ tonalities. It divides flavours into major categories, fruits, brown and bakery, vegetables, meat and seafood, herbs and spices, spirits, and fantasy. Each category is further subdivided into relevant classes. Fruits, as an example, cover citrus, berries, stone fruits, pip fruits and tropical fruits. Languages such as acid, aldehydic, green, floral, juicy, fresh, oily, peely and others apply to each type of fruit.

Reasons for adding flavours can be several, as given below.

1. To offer crops/foods consistent quality

Crop aroma changes after harvest. Further processing generates more changes. Adding flavour to maintain its quality is essential for keeping the good's commercial value.

2. To cover off-notes

Crops can be fractionated into different food ingredients. These ingredients are used to formulate ready-to-eat foods and drinks. These fractions often carry poor aroma or lingering medicinal taste. Adding flavours to mask off-notes improves the product acceptance.

3. To replace costly ingredients. Compounds obtained naturally can be very costly. Artificially synthesised or made is much less expensive hence good alternatives for cost control and supply stability.
4. To replace raw materials which are difficult to handle/source.
5. To drive consumer preferences. All foods should smell and taste good to sell well.

9. Ingredients for Flavouring Creation

Flavours can be applied in beverage, confectionery, dairy, bakery, savory processed foods, snacks, oral care products, pet foods, tobacco, and many other product formulations. Ingredients for creating these flavours can be from natural origin such as essential oils, oleoresins, extracts, juices that are extracted, distilled, expressed or bio-transformed from animal, plant, flower, fruit, and seed source. They can be aroma chemical compounds synthetically made. Reaction flavours using protein and sugar interaction imitating cooking process or smoke flavours from burning woods condensing volatiles to generate appealing savory aroma are widely used in processed foods and ready meals. Various herbs and spices and their blends are applied widely in making foods like snacks, sauces, and meat products. These flavour raw materials and flavours are highly concentrated. Diluting them with food grade solvents or carriers are often necessary. A thousand-fold (1/1000) dilution is normally a starting point.

10. Flavour Labelling

Consumers often label flavours added to food products as chemical and artificial. Flavour is essentially a blend of chemical molecules, regardless of it is naturally formed or synthetically made. It is the same concept as the human body being made up of chemical molecules starting with water.

Flavour molecules can be classified into natural, artificial, or nature-identical (NI) type. A molecule structure existing in nature is the natural type. An example is vanillin that is present naturally in vanilla beans and can be extracted as-is. This vanillin can also be made by chemical synthesis. The resulting compound has the same exact chemical structure as the natural one, it can be classified as natural-identical type. Ethyl vanillin's chemical structure has an ethoxy rather than the methoxy group vanillin has. The structure with ethoxy group does not exist in nature. It is classified as an artificial type. Ethyl vanillin exhibits more powerful vanilla aroma and is more economical to use than vanillin. Since food flavour compounds commercially available nowadays have passed safety screening for human consumption, to minimise consumer resentment on labelling "artificial flavour", China, EU and some other countries have moved away from labelling flavours as artificial or nature identical. All flavours are labelled as "edible flavour or flavouring substances". The US, however, keeps the natural and artificial classification.

Question often arises as to why natural flavour directly extracted or produced from crops cannot be used to flavour drinks or foods? The answers are given below.

1. High cost and low yield

It will take 10,000 kg fresh strawberries to obtain about 1 kg natural strawberry flavour. The 1 kg flavour obtained may flavour about 1000 kg or less products. To extract flavour from the complex fruit texture is challenging. Resulting flavour may not be identical to what exists in the intact fruit because of extraction process induced changes.

2. Availability

Crops can be seasonal. Stable supply at similar maturity level can be an issue. Fruits with different maturity level carries different flavour components.

3. Regulation compliance

Extraction may be limited to restrictions on process and material use requirements.

4. Performance proven to be not optimum

Naturally extracted flavour does not always perform in practical applications from empirical experiences.

5. Purification may not be complete

It can carry undetectable natural toxins.

11. Future of the Flavour Industry

Global flavour trend is going towards minimising animal-origin ingredients, natural, clean label, and green direction for health, wellness, and environmental concerns. Many companies are searching and using ingredients that can be claimed natural and can provide health and wellness function.

To generate more natural raw materials, biotechnology is utilised more to produce via microbes, genetic engineering, and fermentation. Flavour companies use industrial biotechnology to scale up production, to reduce cost, and increase quantity to obtain raw materials in more economical ways. Renewable, eco- and fuel-efficient chemical producing processes are constantly pursued. Major flavour companies follow the 12 principles of green chemistry (American Chemical Society ACS, www.acs.org/greenchemistry) to design and improve their materials, products, processes, and systems to reduce environmental impact and minimise the potential negative health effects of chemicals and chemical synthesis.

Flavour companies possess large databases which include thousands of chemicals — their taste, smell, and application information, formulation combinations, processing changes, production methods, technologies, and others. AI technologies have been increasingly utilised by major players for new innovations. AI technology can shorten trial and error time in creation to deliver products faster. It also provides opportunity for unlimited creativity which may be beyond imagination by human brains within limited time frame to answer customer's request.

12. From Crop to Flavours

Flavours arise from crops. Crops naturally are versatile in food applications amid the global trend of meat reduced, plant-based diet. Crops can be improved to give more variety variations, better quality, and increased quantity. Modern genomics and genetic engineering help to identify genes to grow new variety for obtaining better flavour, keeping quality and endurance to environmental challenges. New natural flavour creations prompt more processed foods diversities and varieties. Combining sciences and biotechnologies, flavour industry can produce more raw materials economically via environmentally friendly ways to cope with increased market demands. With AI technology advancement, more flavour innovations are possible and faster. With all disciplines working and progressing together, more sustainable foods from diversified crops to please all palates can be anticipated.

References

Alonge, M. *et al.*, 2020. Major impacts of widespread structural variation on gene expression and crop improvement in tomato. *Cell*, 182(1): 145–161.e23. https://doi.org/10.1016/j.cell.2020.05.021

BrainFacts/SfN, 2012. Taste and smell, April 1 Article/Newsletter. https://www.brainfacts.org/sensing-thinking-behaving/senses-and-perception/articles/2012/taste-and-smell/\, accessed on October 8, 2020.

Cho, I.H. and Peterson, D.G., 2010. Chemistry of bread aroma: a review. *Food Sci. Biotechnol.*, 19: 575–582.

Flament Ivon, 2002. Coffee flavor chemistry. In: *A Historical Survey of Coffee Aroma Research*, Chapter 4, John Wiley & Sons, Ltd., UK., pp. 53–77.

Heath, H.B., 1978. Description of aromatic profiles — the culinary herbs. In: *Flavor Technology: Profiles, Products, Applications*, AVI Publ. Co., CT., USA., Section I.2, pp.13–25.

ISAAA. Global Status of Commercialized Biotech/GM Crops: 2016. ISAAA Brief No. 52. 2016. (ISAAA: International Service for the Acquisition of Agri-Biotech Applications).

Mouritsen, O.G., 2015. The science of taste. *Flavour Journal-BioMed Central*, 4(18): 1–2. https://doi.org/10.1186/s13411-014-0028-3

Qi Xinpeng *et al.*, 2014. Identification of a novel salt tolerance gene in wild soybean by whole-genome sequencing, *Nat. Comm.* 5, Art #4340. https://doi.org/10.1038/ncomms5340

Reineccius, G., 2006a. Classification of aroma compounds by molecular structure, In *Flavor Chemistry and Technology*, 2nd ed., Chapter 10, CRC Press, pp. 300–303.

Reineccius, G., 2006b. Flavor Analysis, In *Flavor Chemistry and Technology*, 2nd ed., Chapter 3, CRC Press, pp. 33–67.

Scarborough, P., Appleby, P.N., and Key, T.J., 2014. Dietary Green House Gas emission of meat, fish, vegetarian and vegan eaters in UK, Oxford Climate Change Group Research-study in UK.

https://doi.org/10.1142/9789811276736_0002

Chapter 2

Dietary Fiber as a Multifunctional Food Component for Human Health

Peter C.K. Cheung

Food and Nutritional Sciences, School of Life Sciences
The Chinese University of Hong Kong
petercheung@cuhk.edu.hk

1. Definition of Dietary Fiber (DF)

Unlike other food components like proteins, fats, carbohydrates, minerals and vitamins, DF is a concept rather than a well-described chemical entity. Historically from a botanical point of view, DF was regarded as rigid fibrous cell wall constituents which mainly include cellulose and hemicelluloses in plants and chitin in fungi as the structural components. From physiological point of view, DF was firstly defined as remnants of plant cells resistant to hydrolysis by the human alimentary enzymes. This implies that the source of dietary uptake of DF is mainly from plant foods. From chemical point of view, DF was defined as non-starch polysaccharides (NSPs) in plant foods.

With the advancement of recent research on DF, the definition of DF has been modified by various professional and official organizations. In 2010, the Codex Alimentarius Commission from Food and Agriculture Organization (FAO)/World Health Organization (WHO) has defined DF as carbohydrate polymers with ten or more monomeric units/degree of polymerisation (DP), which are non-digestible (not hydrolysed by the endogenous enzymes in humans and animals) and not absorbed in the small intestine. The European Food Safety Authority (EFSA) and US Food and Drug Administration (FDA) have adopted a wider definition of DF to include all carbohydrates that are neither digested nor absorbed in the small intestine and have a DP of three or more monomeric units. They also specify that synthetic and extracted fibers that are not intrinsic to plant cells must also demonstrate physiological effects to human health prior to being declared a dietary fiber. The American Association of Cereal Chemists (AACC) 2003 defines DF as the edible parts of plants or analogous carbohydrates that are resistant to digestion and absorption in the human small intestine with complete or partial fermentation in the large intestine. Under this definition, DF includes polysaccharides, oligosaccharides, lignin and associated plant substances and should promote beneficial physiological effects including laxation and/or blood cholesterol attenuation, and/or blood glucose attenuation.

From food manufacture point of view, the criteria for DF to be used as a functional food ingredient should include resistance to the upper gut tract; beneficial health effects; fermentation by intestinal microbiota; selective stimulation of probiotics; and stability to food processing treatments.

2. DF Components

Figure 1 illustrates the DF components that can be identified in a typical cell wall model in plant cells. With the above definitions of DF, the scope of the components that constitute DF have been expanded. For instance, NSPs have included cellulose, hemicelluloses, arabinoxylans, arabinogalactans, pectins, beta-glucans, inulin, gums such as galactomannans, glucomannans, etc. Other than NSPs, DF components can also include non-digestible oligosaccharides such as oligofructans (FOS) and galacto-oligosaccharides (GOS); analogous carbohydrates including resistant starches, indigestible dextrins, resistant maltodextrins; as well as synthesised carbohydrates including polydextrose, methylcellulose and hydroxypropylmethyl cellulose. Apart from higher

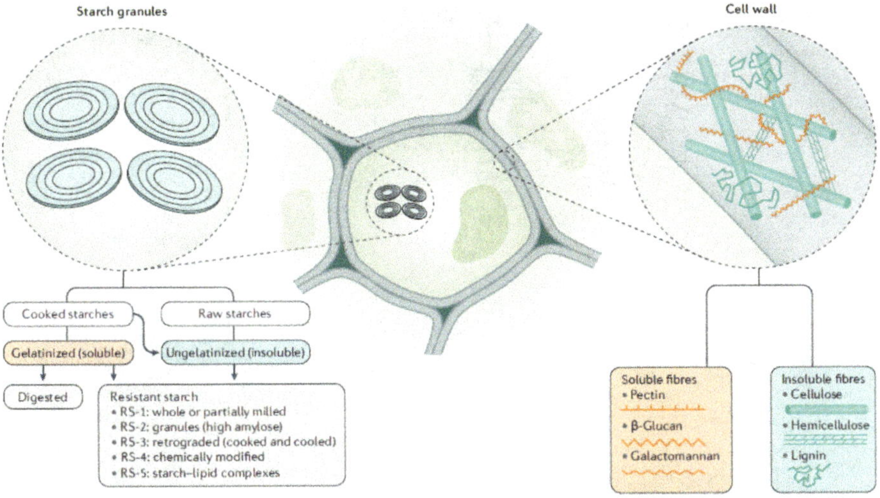

Figure 1. Physicochemical characteristics of DF and their location within the plant cell (Gill *et al.*, 2020).

plants, cell wall components from lower plants including fungi and mushrooms also have NSPs (Figure 2). In particular, edible mushrooms have beta-glucans, mannans and chitin which can be regarded as DF. Non-digestible non-carbohydrates from plants are also considered as DF components. These include lignin and associated substances such as waxes, phytate, cutin, saponins, suberins and tannin.

3. Physicochemical, Functional and Physiological Properties of DF

DF possesses several physicochemical properties that have functional properties which can result in physiological responses in human body with beneficial health effects. Among them, water solubility, viscosity and fermentability are the most important ones that have functional properties such as nutrient bioavailability, transit time of gut digesta along the gastrointestinal (GI) tract, water-holding ability, organic substance adsorption, bacterial degradation, production of microbial metabolites such as short chain fatty acids (SCFAs) and stool weight. For example, DF such

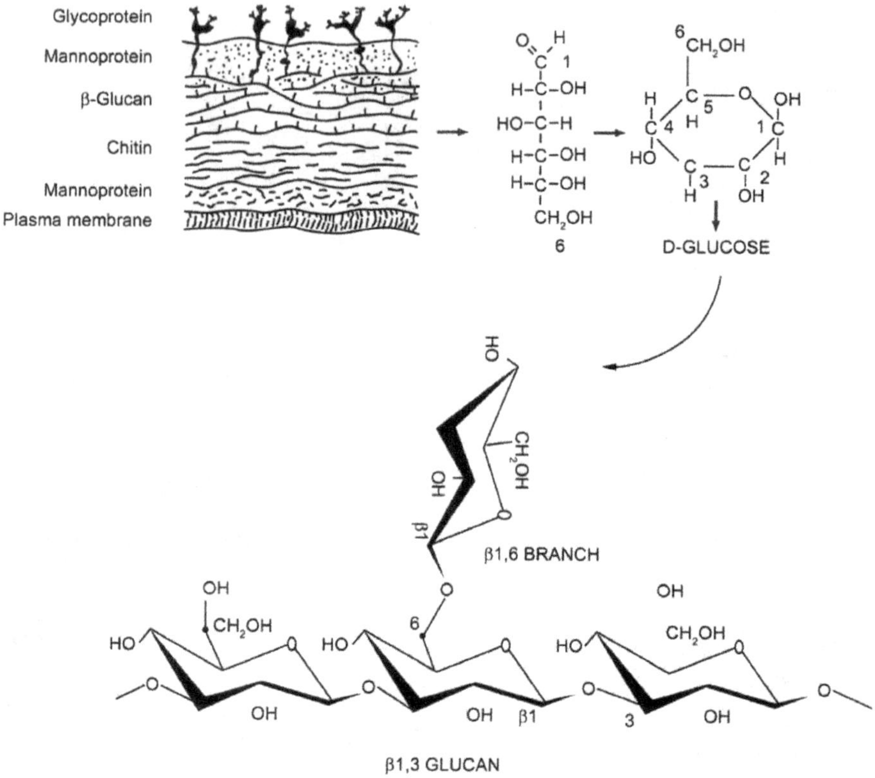

Figure 2. Typical fungal cell wall components with the chemical structure of beta-glucan. (Chan, G. C.F., Chan, W.K. and Sze, D.M.Y., 2009. The effects of beta-glucan on human immune and cancer cells. *J. Hematol. Oncol.* 2, 25. https://doi.org/10.1186/1756-8722-2-25)

as pectin with high water solubility, high viscosity and highly fermentable can reduce nutrient bioavailability and increase production of SCFAs, while DF such as inulin and GOS with high water solubility and fermentability but is non-viscous can only increase SCFA production. DF such as cellulose which is water-insoluble, non-viscous and non-fermentable can induce faster transit time in the GI tract and increase stool weight, while DF such as psyllium with only moderate water solubility and viscosity but low fermentability can have high water holding ability, faster GI transit time and decrease of nutrient bioavailability.

These functional properties would trigger physiological responses that have health implications to human health and diseases. Table 1 lists

Table 1. Physiological Responses affected by the functional properties of DF components.

Functional property	DF fraction	Physiological response
Bacterial degradation	Polysaccharides	Production of short-chain fatty acids, flatulence and acidity
Water-holding capacity	Polysaccharides with polar groups	Effect on nutrient absorption, fecal weight, and rate of transit in stomach and small intestine
Adsorption of organic materials	Lignin Pectin	Binding and excretion of bile acids/neutral sterols
Cation exchange	Acidic polysaccharides	Increase in mineral excretion

Table 2. Characteristics of DF that contribute to improved large bowel function.

Functional property	Physiological response
Water holding capacity	Creates an aqueous phase in the fiber matrix for penetration of microorganisms to break down the polysaccharide structure
Bulk	Provides more material entering the large intestine which serves as substrate for microflora or undigested residue for stool
Fermentability	Provides polysaccharides that can be used by the large intestine microflora and results in increased microbial mass and production of CO_2, H_2, CH_4 and short-chain fatty acids

out some of the functional properties of DF and the related physiological response in our body.

The health benefits that are resulted from the physiological responses due to DF are very wide ranged and have important implications to human health and diseases. For instance, Table 2 shows how the large bowel function can be improved by the functional properties of DF and Table 3 indicates that the lowering of blood cholesterol level and attenuation of blood glucose responses are related to the functional properties of DF. Figure 3 illustrates a detailed proposed mechanism of how the various functional properties of DF can elicit physiological responses that affect GI functions.

Table 3. Characteristics of DF that contribute to the lowering of plasma total and LDL-cholesterol and blunting of glycemic response.

Functional property	Physiological response
Viscosity	Slows digestion and absorption of lipid and carbohydrate Delays gastric emptying Enhances release of cholecystokinin
Water holding capacity	Expands the aqueous phase of the small intestine content
Bulk	Expands bulk phase of intestinal content
Bind bile acids	Increases bile acid excretion-primarily important for cholesterol response

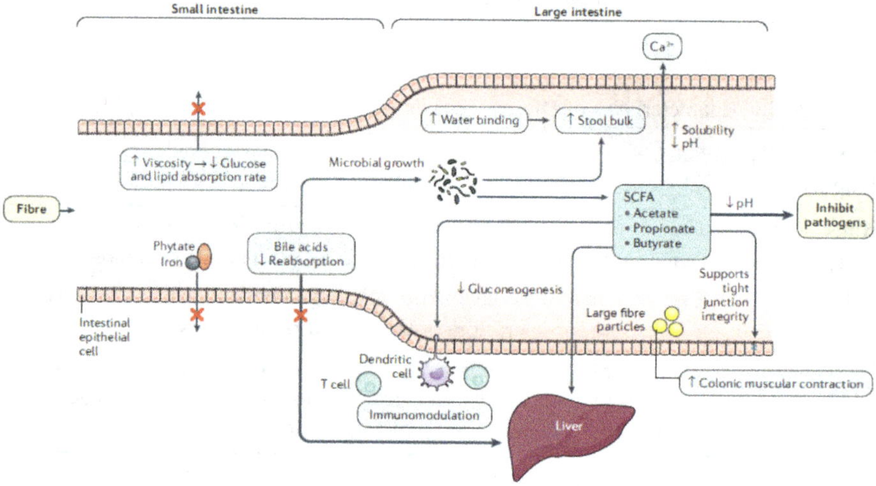

Figure 3. Mechanisms by which DF affect the gastrointestinal tract (Gill *et al.*, 2020).

The question of how the physiological functions of DF can alleviate the harmful effects of human diseases and maintain our body health have been investigated extensively. Figure 4 provides an overview of the plausible mechanisms of the protective effects of DF involved in several chronic diseases including diabetes, coronary heart diseases and cancer.

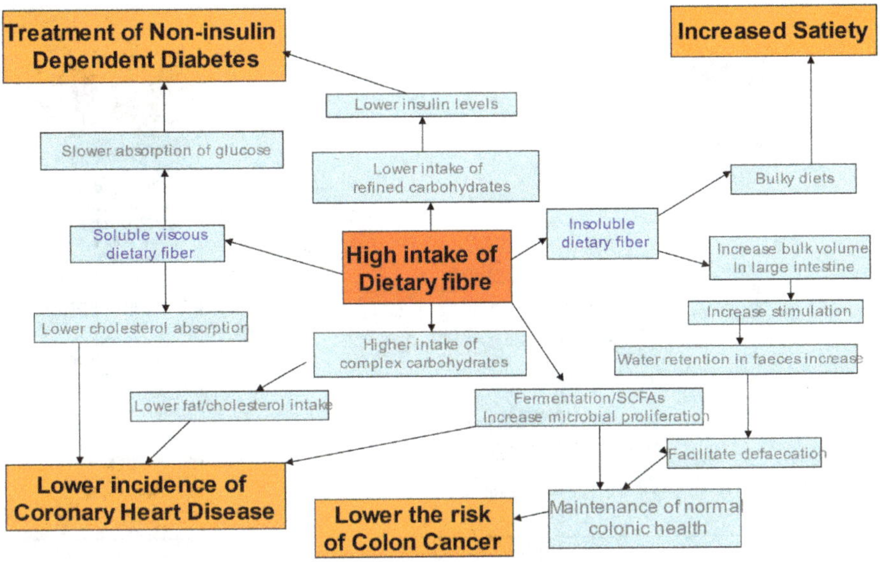

Figure 4. Overview of the plausible mechanisms of the physiological functions of DF.

4. Health Claims and Dietary Intake of DF

Although several health benefits of DF have been identified by research studies, the authorised health claims for DF that meet significant scientific agreement are still limited.

As reference, there are two such health claims made under the US Food and Drug Administration (FDA) Code of Federal Regulations (CFR) Title 21.

- 1st claim: Diets low in fat and high in fiber-containing grain products, fruits and vegetables may/might reduce the risk of some cancers (Section 101.76)
- 2nd claim: Diets low in saturated fat (<10% calories) and cholesterol and high in fruits, vegetables and grain products that contain fiber may/might reduce the risk of coronary heart disease (Section 101.77)

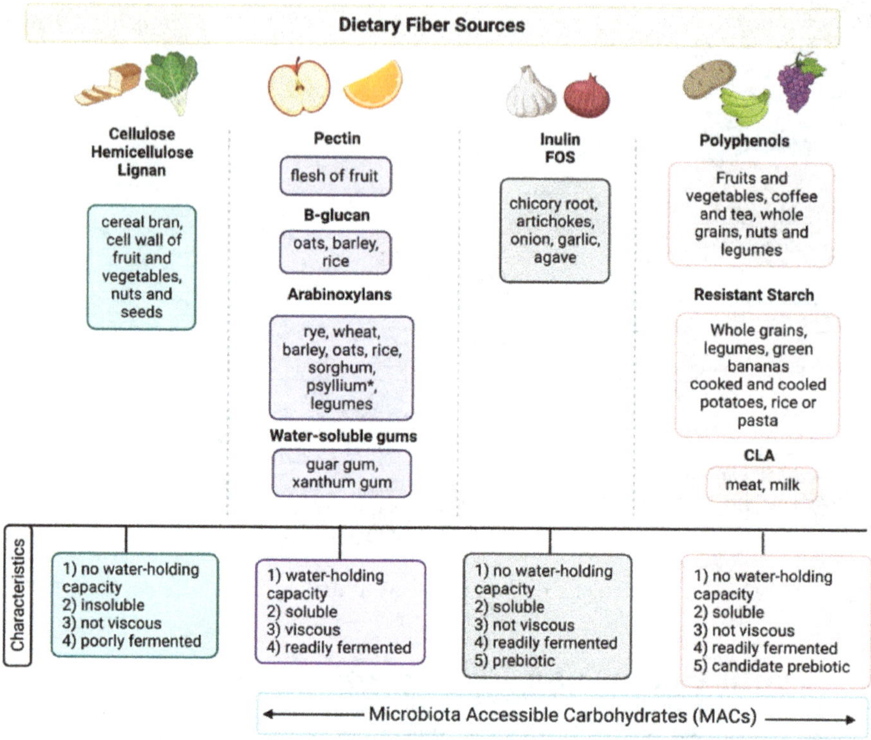

Figure 5. Sources of dietary fiber in the diet and classification based on physical characteristics.

(Haskey, N., Gold, S.L., Faith, J.J. and Raman, M., 2023. To fiber or not to fiber: The swinging pendulum of fiber supplementation in patients with inflammatory bowel disease. *Nutrients*, 15, 1080. https:// doi.org/10.3390/nu15051080)

The daily dietary intake of DF has an important implication on chronic diseases in humans. Majority of countries recommend a daily DF intake for adults of **25–35** g per day (of which 6 g are soluble DF) with average intake of DF under **20** g per day.

Figure 5 shows the major sources of dietary fiber and the classification of dietary fiber that is based on physical characteristics including water-holding capacity, viscosity, and fermentation properties.

Table 4. Impacts of DF intake between 25–29 g/day to the Relative Risk/Risk Ratio of mortality and diseases.

	Relative Risk/Risk Ratio (RR)*	**Confidence Interval (CI)#**
(i) RR of mortality:		
All-cause mortality	0.85	0.79–0.91
Mortality from coronary heart disease	0.69	0.60–0.81
Cancer	0.87	0.79–0.95
(ii) RR of lower incidence of chronic diseases:		
Coronary heart disease	0.76	0.69–0.83
Stroke	0.78	0.69–0.88
Type 2 diabetes mellitus	0.84	0.78–0.90
Colorectal cancer	0.84	0.78–0.89

*RR is the ratio of the probability of an outcome in an exposed group to the probability of an outcome in an unexposed/control group. A RR > 1, then the event is more likely to occur if there was exposure. If RR < 1, then the event is less likely to occur if there was exposure.
#CI is a range of values, above and below the mean, in which the actual value is likely to fall. It represents the accuracy or precision of an estimate. 95% CI of RR implies that one is 95% confident that the RR is between the range indicated.

Apart of individual research projects on DF, meta-analysis of epidemiological cohort studies has been used to evaluate the health effects of DF. Table 4 shows the findings of one of such recent meta-analysis on intake of DF between 25–29 g/day (Reynolds et al., 2019).

In general, DF can reduce mortality and lower the incidence of some chronic diseases as evident from an RR less than 1.

5. Conclusions

Although DF is a non-nutrient component in food, its chemical and physical properties play an important role in their physiological functions that have health implications to humans. More clinical trials and epidemiological data are required to support more evidence-based health claims of DF.

References

Gill, S. *et al.*, 2020. Dietary fiber in gastrointestinal health and disease. *Nat. Rev. Gastroenterol. Hepatol.*, 18: 101–116. https://doi.org/10.1038/s41575-020-00375-4

Kumar, V. *et al.*, 2013. Dietary roles of non-starch polysaccharides in human nutrition: a review. *Crit. Rev. Food Sci. Nutr.*, 52: 899–935. https://doi.org/10.1080/10408398.2010.512671

Reynolds, A. *et al.*, 2019. Carbohydrate quality and human health: a series of systematic reviews a meta-analysis. *Lancet* 393: 434–445. https://doi.org/10.1016/S0140-6736(18)31809-9

Zhang, Y., You, L., Cheung, P.C.K., Liao, L., and Zhao, Z. (2021). Behavior of non-digestible polysaccharides in gastrointestinal tract: a mechanistic review of its anti-obesity effect. *eFood*, 2: 59–72. https://doi.org/10.2991/efood.k.210310.001

https://doi.org/10.1142/9789811276736_0003

Chapter 3

Bioactive Compounds in Crops and Their Impacts on Human Health

Fung Ping Leung* and Wing Tak Wong†

School of Life Sciences, The Chinese University of Hong Kong; State Key Laboratory of Agrobiotechnology, The Chinese University of Hong Kong
**christina_leungfp@cuhk.edu.hk*
†jack_wong@cuhk.edu.hk

Components of food that, when consumed by humans or animals, have an effect on the physiological or cellular processes of those organisms are known as bioactive compounds. Flavonoids, anthocyanins, tannins, betalains, carotenoids, plant sterols and glucosinolates are some of the compounds that fall under this category. They are mostly found in fruits and vegetables, have anti-inflammatory, anti-carcinogenic and antioxidant activities, and can be protective against a variety of chronic diseases such as cardiovascular diseases and cancers. In this book chapter, we discuss the impact of some bioactive compounds (blueberry anthocyanin, lycopene, soy isoflavones and carotenoids) in cardiovascular diseases and cancers, and we highlight their potential mechanisms. Numerous studies, including epidemiological investigations and preclinical studies, have accumulated a mountain of evidence supporting the hypothesis that consuming foods high in the antioxidant flavonoid anthocyanin, like blueberries, can significantly lower one's risk of developing cardiovascular disease. Additionally, we provide compelling evidence showed that lycopene exerts a favourable effect in atherosclerosis, myocardial infarction and hypertension. We also give a succinct overview of the clinical studies and laboratory studies using *in vitro* and *in vivo* models that have demonstrated the effects of isoflavones (genistein and daidzein), two active compounds derived from soybeans, on postmenopause, metabolic syndrome and atherosclerosis as well as their effects on adipocytes, endothelial cells, and vascular smooth muscle cells. Lastly, this book chapter examines the benefits of carotenoids (lutein, α-carotene, β-carotene, zeaxanthin, and lycopene) in mitigating breast and prostate cancers and their possible underlying mechanisms.

Abbreviations

ACh	acetylcholine
ACE	angiotensin converting enzyme
ApoB	apolipoprotein B
AR	androgen receptor
AST/ALT	enzymes aspartate transaminase/alanine transaminase
CAD	coronary artery disease
CK-MB	creatine kinase-MB
CAT	catalase
COX	cyclooxygenase
CK-MB	creatine kinase-MB
CRPC	castration-resistant prostate cancer
CVD	cardiovascular diseases
ER-	estrogen receptor negative
ER	endoplasmic reticulum
eNOS	endothelial nitric oxide synthase
HbA1c	hemoglobin A1c
HDL	high density cholesterol

HDLPD	high density cholesterol particle density
MnSOD	manganese superoxide dismutase
GSH-Px	Plasma glutathione peroxidase
HES1	hairy and enhancer of split-1
HOMA-IR	homeostatic model assessment for insulin resistance
HPLC-MS	high performance liquid chromatography-mass spectrometry
hsCRP	high-sensitivity C-reactive protein
HIF-1	hypoxia-inducible factor-1
HF	high fat
H/R	hypoxia/reoxygenation
HRT	hormone replacement therapy
HV	healthy volunteers
HUVEC	human umbilical vein endothelial cells
IGF-I	insulin-like growth factor I
IL-6	interleukin 6
I/R	ischemia-reperfusion
IPoC	ischemic postconditioning
ISO	isoproterenol
KEGG	Kyoto Encyclopedia of Genes and Genomes
LDH	lactate dehydrogenase
LF	low fat
LWB	lowbush wild blueberries
MPTP	mitochondrial permeability transition pore
NrF2/ARE	nuclear factor erythroid-2-related factor 2/antioxidant responsive element
NHW	non-hispanic white
NHB	non-hispanic black
NO	nitric oxide
8-OHdG	8-hydroxy-2′-deoxyguanosine
PSA	prostate-specific antigen
PrEC	normal human prostate epithelial cells

Phe	phenylephrine
RCTs	randomised controlled trials
RISK	reperfusion injury salvage kinase
ROS	reactive oxygen species
Sirt1	silent mating type information regulation 2 homolog 1
SBP	systolic blood pressure
TC	total cholesterol
TNF	tumour necrosis factor
TNBC	triple negative breast cancer cells
VSMC	vascular smooth muscle cells
VLDL	very low-density lipoprotein
FMD	flow mediated dilation

1. Introduction

1.1. *Bioactive compounds in crops*

Bioactive compounds from crops have gained scientific interest due to their numerous biotechnological applications. They are defined as secondary metabolites that exhibit toxicological or pharmacological effects in man and animals (Shoker, 2020). Bioactive compounds in plants include terpenoids, alkaloids, flavonoids, coumarins and phenolics. They have been commercially used to treat and prevent the risk of several diseases. In plants, bioactive compounds provide protection against biotic and abiotic stress. In humans, they have been extensively used to treat several ailments by generations of indigenous practitioners. These compounds possess unlimited opportunities for new drug discovery because of unmatched availability of chemical diversity. They exhibit a wide spectrum of bioactivities such as immunostimulatory, anti-inflammatory, antioxidant, antimicrobial and anti-cancer effects (Shoker, 2020). Many diseases and conditions are being examined for treatment because of their potential efficacy, safety, and low side effects. Various bioactive compounds will be discussed in this book chapter, along with their uses in the treatment and prevention of cardiovascular illnesses and malignancies.

2. Classification of Bioactive Compounds

The major class of bioactive compounds are (1) terpenes and terpenoids (approximately 25,000 types), (2) alkaloids (about 12,000 types), and (3) phenolic compounds (about 8000 types).

2.1. *Phenolic compounds*

Lignans, flavonoids, stilbenes, and phenolic acids are all examples of these compounds. The most numerous and diverse class of phenolic chemicals are called flavonoids. Flavanones, flavanols, anthocyanins, flavones, and isoflavones are all examples of these compounds. Stilbenes are another type of compound, and they are present in a wide variety of food plants, including peanuts and berries (Shoker, 2020). Phenolic acids are another class of phenolic compounds. Phenolic acids are abundant in coffee, fruits (e.g. apple, berries and orange), vegetables (e.g. kale, brussel sprouts and broccoli) and wholegrains. They include vanillic, gallic, salicylic acids, coumaric and caffeic acids. The last class consists of phytoestrogens known as lignans. They are found in higher concentrations in sesame and flax seeds.

2.2. *Alkaloids*

They are organic nitrogen compounds generated from plants that include bases. There are morphine, caffeine, and nicotine among them. Morphine is a potent analgesic used to alleviate pain. Nicotine derived from the tobacco plant is the most addictive substance in cigarettes. (Shoker, 2020). Caffeine is a purine alkaloid that is mostly found in coffee seeds. It is a natural pesticide that paralyses and kills most insects feeding on plants.

2.3. *Terpenes and terpenoids*

Terpenes are straightforward hydrocarbons that are assembled from isoprene molecules. Terpenoids, on the other hand, belong to a subclass of terpenes that has been transformed by the addition of oxidised methyl groups that have been rearranged or eliminated from various locations

(Shoker, 2020). They are classified into different groups depending on the number of isoprene units. They include monoterpenes, triterpenes, tetraterpenes, and polyterpenes.

3. Role of Different Bioactive Compounds in the Prevention of Cardiovascular Diseases

Cardiovascular diseases (CVDs) are a serious health concern around the globe and the prime cause of many deaths. According to World Health Organization (WHO), CVDs will be the major cause of 23.3 million deaths around the world by the year 2030 (Shoker, 2020). The categories of CVD include coronary artery disease (CAD), hypertensive heart diseases, angina, and stroke. Atherosclerosis (hardening of the arteries) is the basis of several CVDs such as stroke and CAD. It is characterised by deposition of cholesterol and fat inside the arteries resulting in the blockage of coronary arteries. It is a multifactorial disease with a range of various risk factors. These factors include obesity, hypertension, high cholesterol levels, type 2 diabetes, menopause and unhealthy diet. Menopausal women are at a high risk of getting CVDs because of low estrogen levels (Dosi *et al.*, 2014). Reduced levels of this hormone are linked to high blood pressure and elevated low-density lipoprotein (LDL) cholesterol levels that result in atherosclerosis. Transitioning to a healthier, varied, and balanced diet is crucial in preventing CVDs and ensuring good health and quality of life. Experts recommend taking more plant products such as fruits and vegetables and fewer animal-based diet such as meat (Mulero *et al.*, 2015). Bioactive compounds are found in crops and they provide extra-nutritional constituents that are essential in preventing CVDs.

3.1. *Blueberry anthocyanins*

3.1.1. *Blueberry anthocyanins and clinical studies in cardiovascular diseases*

Blueberry consumption has been shown to be beneficial in the prevention of a variety of chronic illnesses, including CVD. A growing body of human clinical studies have explored the potential positive health benefits of blueberries in healthy individual and those with type 2 diabetes, or postmenopausal women with pre- or stage 1-hypertension or metabolic syndrome (Johnson *et al.*, 2015). A randomised, controlled, double-blind, crossover intervention trial established that consumption of blueberries enhanced endothelial function in healthy males time- and dose-dependently. These benefits may be mediated by the influence of circulating phenolic compounds (vanillic acid, hippuric acid and homovanillic acid) on neutrophil NADPH oxidase activity (Rodriguez-Mateos *et al.*, 2013). In addition, Rogers-Mateos *et al.* (2019) investigated the impact of blueberry anthocyanins and circulating metabolites in causing changes in vascular function. By flow-mediated dilation, purified anthocyanins enhanced endothelial function in healthy persons. Daily consumption of wild blueberries for a month increased flow-mediated dilatation and lowered ambulatory systolic blood pressure. Fourteen (acute flow-mediated dilatation)and twenty one (chronic flow-mediated dilatation) of the 63 anthocyanin plasma metabolites connected to acute and chronic flow-mediated dilatation respectively. The administration of these metabolites also improved flow-mediated dilatation in mice. Six hundred and eight genes and three microRNAs, including miR-181c, miR-126, and miR-30c, showed significant changes in differential expressions when wild blueberries were consumed on a daily basis in people. Cell adhesion, migration, immunological response, and differentiation were affected in the pathway enrichment analysis (Rodriguez-Mateos *et al.*, 2019). Further, a long-term randomised controlled trials (RCTs) has shown that consuming 1 cup of blueberries daily for six months improved endothelial function, reduced systemic arterial stiffness, and lowered cyclic guanosine monophosphate concentrations. As a result of the 1-cup/day intervention and the addition of apolipoprotein A-I, high-density lipoprotein cholesterol (HDL) and HDL particle density (HDLPD) were both increased in non-statin users following the intervention (Curtis *et al.*, 2019).

Stote and colleagues in 2020 conducted the first study to examine the health benefits of consuming 22 g freeze-dried blueberries (about 1 cup of fresh blueberries) daily for eight weeks in patients with type 2 diabetes. Individuals who consumed freeze-dried blueberries had significantly reduced levels of glycemic control biomarkers such as hemoglobin A1c

(HbA1c) and fructosamine (Stote *et al.*, 2020). Recently, a double blind RCT in metabolic syndrome participants demonstrated that a single dosage of freeze-dried powdered blueberries (equal to 1 cup of fresh blueberries) reduced the negative effects of a high-fat, high-sugar meal consumed within 24 hours in an at-risk population while concurrently reducing blood sugar, cholesterol, and apolipoprotein A1 (ApoA1) fractions (Curtis *et al.*, 2022).

Hypertension and arterial stiffness are more common in postmenopausal women, raising their risk of cardiovascular disease. There are antihypertensive medication therapy, but natural remedies are becoming more popular. Johnson *et al.* (2015) found a significant group-time interaction for systolic and diastolic blood pressure in the blueberry powder group, but not in the control powder group. After eight weeks, the blueberry powder group had significantly higher levels of nitric oxide than the control group. Blueberry eating on a daily basis may lower blood pressure and arterial stiffness, perhaps due to enhanced nitric oxide (NO) synthesis (Johnson *et al.*, 2015).

In postmenopausal women with pre- or stage 1-hypertension, blueberries' effects on oxidative stress, inflammation, and antioxidant defense were studied. Forty pre- and postmenopausal women aged 45 to 65 were given either 22 g freeze-dried highbush blueberry powder (Blueberry) or 22 g of a placebo powder for an eight-week period (Control). In this study, 8-hydroxy-2′-deoxyguanosine (8-OHdG, a blood biomarker of oxidative DNA damage) was measured at the beginning of the study, along with oxidative stress, inflammation, and antioxidant defences. The levels of 8-OHdG in Blueberry group were considerably lower than in Control group after 4 weeks, indicating a time-treatment interaction. After eight weeks, there had been no improvement from 0 week. These data indicated that daily blueberry eating was initially protective against oxidative DNA damage, however, it was unable to maintain this benefit over a longer period of time (Johnson *et al.*, 2017).

3.1.2. *Animals studies in cardiovascular diseases*

Wild blueberries in the diet have been shown to alter the contractile mechanism of vascular smooth muscle by reducing alpha1-adrenergic receptor agonist-mediated contraction without affecting membrane sensitivity of endothelial or vascular smooth muscle cell layer (Norton *et al.*, 2005). In addition, the aortas of weanling male Sprague-Dawley rats were examined for a period of seven weeks while they were either fed a control diet or a diet consisting of blueberries. The Blueberry diet reduced the vasoconstrictor response to phenylephrine (Phe) in rat aortic rings. Inhibition of NO synthase but not cyclooxygenase (COX) increased the constrictor response in both dietary groups, but more in the Blueberry group. Similarly, the NO pathway was more involved in ACh-induced endothelium-dependent vasorelaxation in Blueberry-fed rats. COX inhibition had no influence on maximal ACh-induced vasorelaxation in any diet group. Aortic rings in Blueberry group were less sensitive to vasoconstrictor and vasodilator than controls (Kalea *et al.*, 2009).

Bacteria in the gut play a crucial role in metabolic and immunological processes. Infection-causing bacteria can be impacted by dietary factors. The addition of lowbush wild blueberries (LWB) to the diet was looked at as a way to improve gut health. For 6 weeks, Sprague Dawley rats were fed either a control diet (AIN93) or a LWB-enriched diet (AIN93 with 8% LWB powder added in place of dextrose). With regard to 3 phyla and 22 taxa, the research conducted by Lacombe A *et al.* (2013) demonstrated a change in relative abundance. When LWB was added to the diet, Lactobacillus and Enterococcus became less common. Hierarchical analysis shows that there was also a big rise in the relative number of Actinobacteria, Actinomycetales, and many new genera in the Bifidobacteriaceae and Coriobacteriaceae families. The LWB-diet appeared to affect around 9% of the 4709 Kyoto Encyclopedia of Genes and Genomes (KEGG) hits discovered by functional annotation of Illumina shotgun sequencing and bioinformatics. The number of Open Reading Frames (ORFs) allocated to the KEGG category xenobiotics (pollutants) biodegradation and metabolism was considerably higher in the LWB-enriched diet compared to the control, and this included the route for benzoate breakdown (Lacombe *et al.*, 2013).

Obesity-related chronic inflammation has been linked to gut microbial dysbiosis. Obesity-related comorbidities may thus be affected by microbiota alteration (Cuevas-Sierra *et al.*, 2019). As a source of anti-inflammatory anthocyanins, blueberries have the potential to change our gut microbiome for the better. Blueberry increased Gammaproteobacteria abundance compared to Low Fat (LF) and High fat (HF) rats. Ileal villus height decreased by 15% in HF rats, but was recovered by blueberry supplementation. Mucin2 expression in the ileum was 150 percent greater in HF Blueberry rats than in HF rats, but expression in the LF group was similar to that in the HF and HF Blueberry groups. Proinflammatory cytokine (TNFα and IL1β) gene expressions in visceral fat were enhanced by HF diet compared to LF feeding significantly and adjusted by blueberry supplementation. Additionally, blueberries enhanced insulin sensitivity (Lee *et al.*, 2018). The effects of blueberries on gut microbiota and phenolic metabolism to bioactive metabolites are unclear. Recent study by Cladis *et al.* (2021), 5-month-old ovariectomised Sprague-Dawley rats were gavaged for 90 days with pure blueberry polyphenol extract (0, 50, 250, or 1000 mg total polyphenols/kg body weight) or blueberries (50 mg total polyphenols/kg body weight, corresponding to 250 g fresh blueberries in humans). Moderate dosages (50 and 250 mg) enhanced gut microbial diversity whereas high doses diminished it. The majority of urinary phenolic compounds were microbially generated and underwent significant host xenobiotic phase II metabolism. Further, gut bacteria are able to metabolise polyphenols and other substances found in wild blueberries, which may have a direct effect on the intestinal environment (Cladis *et al.*, 2021).

A randomised, double-blind, placebo-controlled trial of six weeks of wild blueberry (*Vaccinium angustifolium*) drink intake against a placebo drink was conducted to examine the influence on the intestinal flora of human volunteers. Bifidobacterium spp. and *Lactobacillus acidophilus* grew considerably after blueberry treatment compared to total eubacteria. Bacteroides species, Prevotella species, Enterococcus species, and Clostridium coccoides species did not show any significant changes. The polyphenols and fiber found in wild blueberries may play a key role in the beneficial effects of bifidobacteria on the host, as has been widely hypothesised. A wild blueberry drink may have a good effect on the makeup of the gut flora when consumed on a regular basis (Vendrame *et al.*, 2011).

On the whole, the consumption of blueberries have good impacts on cardiovascular health by: (i) improved endothelial function in time and dose-dependent manner; (ii) increased flow-mediated dilatation; (iii) decreased systemic arterial stiffness; (iv) decreased cyclic guanosine monophosphate concentration; (v) increased of ApoA1, high density lipoprotein cholesterol and HDL particle density; (vi) reduced levels of glycemic control biomarkers like HbA1c and fructosamine; (vii) reduced blood sugar and cholesterol; (viii) lower blood pressure and arterial stiffness as a result of nitric oxide synthesis; (ix) protective against oxidative DNA damage in DNA; (x) blueberry polyphenols in food and supplements change gut microbial populations and phenolic metabolism, which could potentially exert health benefits.

3.2. *Lycopene*

3.2.1. *Lycopene and clinical studies in cardiovascular diseases (atherosclerosis, myocardial infarction and hypertension)*

Among the major terpenoids, lycopene is the most powerful antioxidant that plays an important role in preventing CVDs in humans. This compound has conjugated double bonds that give antioxidant activity that enables it to protect proteins, DNA and lipids against oxidation. Fruits and vegetables that are red in colour, such as tomatoes, papayas, and watermelons, are good sources of this pigment. Approximately 90% of the total carotenoids that make up the tomatoes are composed of lycopene (Petyaev, 2016). Based on a study conducted by Framingham Heart, lycopene consumption is associated with reduced incidences of CVDs such as angina pectoris, myocardial infarction, and coronary disease (Petyaev, 2016). Extensive studies also found low plasma levels of lycopene in patients with hypertension, atherosclerosis, and stroke.

The risk of developing coronary artery disease dramatically increases after menopause, a pro-atherogenic condition. This pilot investigation aimed to test LycoRed as an alternative to hormone replacement therapy (HRT) or preventing coronary artery disease in postmenopausal women (containing 2000 g lycopene). Over the course of six months, 41 postmenopausal women in good health were randomly assigned to receive either continuous combination HRT or LycoRed. Total cholesterol was lowered by 23.5%, LDL was lowered by 19.6%, and HDL was raised by

38.9% after 6 months of HRT treatment. The LycoRed group saw similar changes in triglycerides (–24.2%), LDL (–14.9%), and HDL (+26.1%). Both groups' triglycerides climbed at 6 months, but not at 3 months. The very low-density lipoprotein (VLDL) levels of either group barely budged. Glutathione levels rose by 5.9% and 12.5% in the HRT and LycoRed groups, respectively. Indicators of oxidative stress and blood lipids were similarly affected by both HRT and LycoRed. In place of HRT, postmenopausal women can take LycoRed to help reduce their risk of atherosclerosis (Misra *et al.*, 2006).

As mentioned before, lycopene is a potent antioxidant found in such diets with evidence suggesting beneficial effects. The effects of lycopene were examined on the vasculature in CVD patients and separately, in healthy volunteers. In a double-blind experiment, the researchers assigned 36 statin-treated CVD patients and 36 healthy volunteers (HV) to daily 7 mg lycopene or placebo for 2 months. Using venous plethysmography, they assessed forearm responses to intra-arterial acetylcholine (EDV), sodium nitroprusside (EIDV), and NG-monomethyl-L-arginine (BNA) infusions. Secondary vascular and biochemical endpoints were also investigated. Post-lycopene EDV improved by 53% without affecting EIDV or baseline NO responses. The lycopene had no effect on EDV in HVs. In both the CVD and healthy volunteers arms, lycopene had no effect on blood pressure, arterial stiffness, lipids, or High-sensitivity C-reactive protein (hsCRP). Contrary to expectations, CVD patients had lower LDL cholesterol (1.2 mmol/L) than HV patients at baseline. EDV responses after lycopene treatment were similar to baseline HV responses indicating lycopene enhanced endothelial function. Lycopene enhances endothelial function in CVD patients but not in healthy volunteers (Gajendragadkar *et al.*, 2014).

Oxidative stress and antioxidant deficiency play a crucial role in the onset, progression, and resolution of cardiovascular disease. Heart variables, inflammation markers, and oxidation were studied after giving highly bioavailable lycopene to people with coronary artery disease for 30 days. Patients were randomly assigned to receive either 7 mg of lactolycopene or lycosome-formulated GA lycopene once daily. They studied lipids, cardiovascular function, and four oxidative stress and inflammatory indicators. Lycosome-formulated lycopene raised blood lycopene by 2.9

and 4.3 times after 2 and 4 weeks, whereas lactolycopene increased it by 50%. Inflammatory oxidative damage marker and Chlamydia pneumoniae IgG were reduced by a factor of three. Oxidised LDL was cut in half by taking lycopene in lycosome form. The results showed that at the end of the trial, tissue oxygenation and flow-mediated dilation were improved after supplementation with lycosome-formulated GA lycopene. There was no influence of lactolycopene on the measured properties though. This means that increased bioavailability of lycopene increases its antioxidant, anti-inflammatory, and cardiovascular benefits (Petyaev *et al.*, 2018).

High doses of bioavailable lycopene were administered to people with coronary artery disease for 30 days. Tomatoes (*Solanum lycopersicum*) are well-known for the high levels of lycopene, a carotenoid that promotes healthy lipid metabolism. Researchers in Japan studied the effects of a new high-lycopene tomato, PR-7, on 74 healthy volunteers with low-density lipoprotein-cholesterol (LDL-C) readings between 120 and 160 mg/dL for 12 weeks. Participants were randomly assigned to receive either a high-lycopene tomato supplement or a placebo (lycopene-free tomato). Participants in the high-lycopene group took 50 g of semidried PR-7 (lycopene, 22.0 – 27.8 mg/day) every day for 12 weeks, whereas those in the placebo group consumed semidried tomato. In contrast to the placebo group, this one saw an increase in lycopene. Reduced LDL-C and safety were both seen with the high-lycopene tomato PR-7 (Nishimura *et al.*, 2019).

Diabetic people are at a higher risk of developing cardiovascular disease. Because of hyperglycemia, insulin resistance, and dyslipidemia, as well as inflammation and oxidative stress, atherosclerosis, the common foundation of most cardiovascular problems, is more frequent and widespread in this group. Lycopenes are bioactive compounds with antioxidant and anti-inflammatory properties that are mostly found in tomatoes and tomato derivatives. In cross-sectional research of 105 newly diagnosed diabetic participants, they looked at the relationship between circulating lycopenes and carotid plaque load. Sonographic assessment of carotid arteries was used to determine the presence of atheroma plaque (wall thickness 1.5 mm), the number of plaques, and the plaque load (sum of maximum heights of all plaques). High performance liquid chromatography-mass spectrometry (HPLC-MS) was used to measure plasma lycopenes

(5-cis-, 9-cis-, 13-cis-, and trans-lycopene). Atheroma plaque was found in 75 subjects, 38 of whom had one plaque and 37 had two or more carotid plaques. There were no variations in plasma lycopene concentrations between participants with and without atherosclerotic plaques. Conversely, 5-cis-lycopene, all cis-lycopene isomers, trans-lycopene, and total lycopene isomers were correlated adversely with plaque load. Lycopene levels in the blood are inversely related to atherosclerotic burden. They provide new findings indicating that consuming lycopene found in tomatoes and tomato byproducts may be advantageous in the prevention of atherosclerosis (Chiva-Blanch *et al.*, 2020).

An early study by Kohlmeier *et al.* (1997) reported that lycopene, or another strongly linked component, may contribute to vegetable consumption's preventive impact against myocardial infarction risk. Biopsies of adipose tissue were obtained quickly after an acute heart attack and tested for carotenoids and tocopherol levels (Kohlmeier *et al.*, 1997). There have been a number of epidemiological studies that suggest a negative correlation between cardiovascular events and high dietary consumption or plasma levels of carotenoids (Ruiz Rejón *et al.*, 2002). Having high levels of carotenoids in the blood or adipose tissue, or eating a diet rich in carotenoids, may reduce the risk of a heart attack (Kabagambe *et al.*, 2005; Kardinaal *et al.*, 1993). According to the findings of Karppi and coworkers' study, males who have low blood concentrations of lycopene and beta-carotene may have an increased risk of developing acute myocardial infarction (AMI) (Karppi *et al.*, 2012).

3.2.2. *Lycopene and animal studies in cardiovascular diseases (myocardial infarction)*

Oral lycopene absorption prevents cardiac I/R damage. Due to significant first-pass metabolism, lycopene has little bioavailability and cannot be utilised clinically. Tong *et al.* (2016) produced an *in vitro* hypoxia/reoxygenation (H/R) cell model and an *in vivo* regional cardiac ischemia-reperfusion (I/R) animal model by ligating the left anterior descending artery. Before reoxygenation, lycopene reduced H/R-induced cardiomyocyte death. Intravenous lycopene decreased MI, ROS production, and JNK activation in mouse heart tissue during *in vivo* regional I/R. Intravenous infusion of 1 microgram per milliliter of blood protects mice against cardiac I/R injury by reducing ROS production and inflammation (Tong *et al.*, 2016).

The mitochondrial permeability transition pore (MPTP) is a therapeutic target for myocardial ischemia-reperfusion injury (MIRI). Lycopene protects against MIRI. Pretreatment with lycopene improved cell viability, reduced myocardial infarction size, and reduced apoptosis, according to Li *et al.* (2019). Lycopene also reduced the expression of cytochrome *c*, APAF-1, cleaved caspase-9, and cleaved caspase-3. MPTP opener negated MIRI's benefits. Lycopene treatment increased Bcl-2 while lowering Bax and the Bax/Bcl-2 ratio. They found that lycopene reduced MPTP opening through regulating the Bax and Bcl-2 transcription factors (Li *et al.*, 2019).

Patients with high cholesterol lose the cardioprotective benefits of ischemic postconditioning, which prevents MIRI. Lycopene restored IPoC's cardioprotective effects in hypercholesterolemic rats. A high-cholesterol diet was administered to male Wistar rats for 12 weeks to simulate hypercholesterolemia. Separated rat hearts were ischemia-reperfused for 30 minutes. Animals received intraperitoneal lycopene five days before ischemia and reperfusion. In hypercholesterolemic rats, IPoC and low-dose lycopene reduced infarct size, enzyme release, and cardiomyocyte apoptosis. IPoC inhibited ERK1/2 and GSK3 while phosphorylating AKT. This combination lowered cytochrome C and caspase 9 and 3 cleavage products, which are implicated in mitochondrial membrane permeability transition. They also suggested that lycopene restored IPoC's cardioprotective effects on MIRI in hypercholesterolemic rats by inhibiting ER stress and reactivating the Reperfusion Injury Salvage Kinase (RISK) pathway in the myocardium (Duan *et al.*, 2019).

Additionally, researchers sought to discover if lycopene might stop apoptosis and reduce oxidative stress to prevent cardiac ischemia-induced damage. Rats were separated into five groups: sham, model, low-, medium-, and high-dose of lycopene treatment. Rats given subcutaneous isoproterenol for two days developed myocardial ischemia. Model group CK-MB, TC, and triglycerides (TG) levels were higher. CK-MB, TC, and

TG were reduced in Lycopene-treated groups. Lycopene increased activities of SOD, CAT, GSH, GPx, and Malondialdehyde (MDA) content. Controls had decreased TNFα, IL-6, and IL-1. Medium- and high-dose lycopene reduced IL-1, TNFα, and IL-6. Models had increased Bax/Bcl-2, Cyt-c, and caspase-3. Medium- and high-dose Lycopene restored Bax/Bcl-2, Cyt-C, and Caspase-3. p-JNK/JNK kinase, p-STAT1 (Ser727)/Ser721, p-Ser725, and p-Ser725 expression rose, but p-STAT3 (Tyr705, Ser727)/STAT3 and p-STAT1 (Tyr701)/STAT1 expression dropped. p-JAK and p-STAT1 (Tyr701)/STAT1 expression levels declined in the medium- and high-dose groups, but p-STAT1 (Ser727)/STAT1 and p-STAT3 (Tyr705)/STAT3 expression levels increased with lycopene. Lycopene may reduce JNK/ERK signalling to restore myocardial ischemia-induced oxidative damage and cardiomyocyte death (Fan *et al.*, 2019).

A study carried out by Upaganlawar *et al.* (2010) investigated vitamin E and lycopene's cardioprotective efficacy against isoproterenol-induced myocardial infarction in rats. After injecting rats with isoproterenol (ISO), researchers observed that the medication increased organ weight, lipid peroxidation, serum marker enzymes (AST/ALT, CK-MB, and LDH), and Ca^{2+} ATPase activity, while lowering body weight and endogenous antioxidant activities (GSH, GPx, GST, SOD and CAT). ISO exposure gave LDH1-LDH2 a prominent band. Vitamin E and lycopene over 30 days mitigated these effects compared to individual treatment and ISO. Histopathological results were consistent with the biochemical values. Their results show that when vitamin E and lycopene are given together, they help protect the heart during ISO-induced myocardial infarction in rats (Upaganlawar *et al.*, 2010).

Treating hypertension reduces the risk of CVD. Lycopene, beta carotene, and vitamin E are powerful antioxidants contained in tomato extract that help inactivate free radicals and prevent atherosclerosis. Engelhard *et al.* (2006) reported that a short-term treatment with an antioxidant-rich tomato extract (8-week treatment period with tomato extract 250 mg Lyc-O-Mato) can lower blood pressure in individuals with grade-1 hypertension who have not previously received medication therapy. It is still needed to establish that this therapy has a continuous impact and that it has a long-term positive effect on cardiovascular risk factors (Engelhard *et al.*,

2006). Similarly, another study also demonstrated that tomato extract exhibited a clinically meaningful benefit when administered to individuals receiving low doses of angiotensin converting enzyme (ACE) inhibitors, calcium channel blockers, or their combination with low dosage diuretics, lowering blood pressure by more than 10 mmHg systolic and more than 5 mmHg diastolic pressure (Paran *et al.*, 2009). In hypertensive individuals, results showed that only carotenoid levels of 15 mg lycopene or higher in the tomato nutrient complex dosage are associated with a reduction in systolic blood pressure (SBP); smaller doses and lycopene alone are not (Wolak *et al.*, 2019). In addition, a recent meta-analysis imply that a lycopene supplement of more over 12 mg/day may successfully lower SBP, particularly in Asians or populations with higher baseline SBP (Li and Xu, 2013). Further, those with hypertension were shown to have higher levels of uric acid, whereas those with higher levels of lycopene were found to have lower levels of hypertension. The ratio of blood lycopene to serum uric acid was also shown to be significantly associated with hypertension in people who were overweight or obese (Han and Liu, 2017).

Taking into account epidemiological data, clinical research, and experimental data, lycopene in the human diet has beneficial effects on cardiovascular health by (i) reduced total cholesterol, lowered bad cholesterol LDL, but raising good cholesterol HDL; (ii) decreasing in Chlamydia pneumoniae IgG by threefold which is known to be an important marker of oxidative stress and inflammation in cardiovascular patients with coronary artery diesases; (iii) decreasing in the markers of biological oxidation (oxidised LDL); (iv) increased tissue oxygenation and flow-mediated dilation; (v) associated with reduced atherosclerotic burden/plaque; (vi) reduced hypoxia/reoxygenation (H/R)-induced cardiomyocyte death; (vii) improved cardiac cell viability, decreased myocardial infarction size, and reduced apoptosis by modulating the mitochondrial permeability transition pore (MPTP); (viii) restored ischemic postconditioning (IPoC)'s cardioprotective effects on hypoxia/reoxygenation MIRI in hypercholesterolemic rats by inhibiting ER stress and reactivating the Reperfusion Injury Salvage Kinase (RISK) pathway in the myocardium; (ix) reduced JNK/ERK signalling to decrease the ischemia-induced oxidative damage and cardiomyocyte death and (x) a reduction in systolic blood pressure.

3.3. *Soy isoflavones*

3.3.1. *Soy isoflavones and clinical studies in cardiovascular diseases (postmenopause, metabolic syndrome and atherosclerosis)*

A woman's risk of cardiovascular disease rises after menopause because her body produces less estrogen (Kannel *et al.*, 1976). Clinical and epidemiological evidence suggests that postmenopausal women's health benefits from consuming more soy foods (Messina, 2002). Isoflavone supplementation in postmenopausal women significantly enhanced flow mediated dilation (FMD) levels compared to placebo, according to a meta-analysis of nine studies (Li *et al.*, 2010). It is also reported that soy isoflavones lowered cardiovascular risk in postmenopausal women. The 10-year risk of coronary heart disease was reduced by 27% and the risk of a myocardial infarction was reduced by 37% (Sathyapalan *et al.*, 2018). In postmenopausal women, isolated soy protein dramatically decreased blood levels of TG, TC, LDL-C, HDL-C, and Apo-B (Moradi *et al.*, 2020). Additionally, postmenopausal women may benefit from using soy and its isoflavones, which help rectify alterations in plasma lipid metabolism (Barańska *et al.*, 2021).

Soybean-derived isoflavone has been shown to have estrogenic and possibly cardioprotective properties, as well as improved endothelial dysfunction, in several laboratory investigations. In ovariectomised rats, the elimination of soy isoflavones from the diet lowered endothelium-derived NO levels. S-equol (an intestinal bacterial metabolite of the soybean isoflavone daidzein) supplementation increased NO-related endothelial function to some degree (Ohkura *et al.*, 2015). Endothelial dysfunction produced by estrogen deprivation in rats was studied using genistein, a flavonoid-based phytoestrogen obtained from a soy diet. Genistein supplementation and estrogen replacement therapy improve endothelial dysfunction induced by ovariectomy in rats. Ovariectomy-induced endothelial dysfunction can be improved by treatment with genistein and estrogen replacement therapy (Squadrito *et al.*, 2000).

The metabolic syndrome is a multifaceted disorder characterised by visceral obesity, dyslipidemia, hypertension, and insulin resistance

(Wilson *et al.*, 2005). Elevated blood levels of inflammatory markers such as C-reactive protein (CRP), interleukin (IL)-2, IL-6, IL-18, and tumour necrosis factor (TNF), as well as endothelial dysfunction, are linked to metabolic syndrome symptoms (Koh *et al.*, 2005). Inflammation, insulin resistance, and endothelial function are all tightly linked in people with the metabolic syndrome, and they can all affect and worsen metabolic decline (Esposito and Giugliano, 2004; Esposito *et al.*, 2005). Endothelium-dependent vasodilation diminishes with age, notably in women after menopause (Gerhard *et al.*, 1996). As a result, postmenopausal women with metabolic syndrome are more likely to develop endothelial dysfunction.

In the development of metabolic syndrome, diet plays a crucial role, in part via affecting proinflammatory markers (Grandl and Wolfrum, 2018). Lower levels of inflammation markers and enhanced endothelial function have been linked to the use of soy products (Azadbakht *et al.*, 2007). Isoflavones may be responsible for the reduction in TNFα in post-menopausal women on a soy-based diet (Huang *et al.*, 2005). Beyond inflammation markers and endothelial function, the potential benefits of soy protein and isoflavones on cardiovascular metabolic risk variables such as body weight and adiposity (Velasquez and Bhathena, 2007), blood pressure (Appel, 2003), glycemic control (Vedavanam *et al.*, 1999), and lipid profiles (Torres *et al.*, 2006) have been studied. There is substantial evidence that soy protein/isoflavones protect against metabolic abnormalities. A prospective analysis of the Korean Multi-Rural Communities Cohort Study (MRCohort) study revealed that consuming soy protein and isoflavones on a regular basis is inversely linked with the risk of metabolic syndrome and its components (Woo *et al.*, 2019).

In healthy postmenopausal women, previous research has revealed that genistein may have positive benefits on endothelial function and cardiovascular risk factors. The impact of genistein on endothelial function in postmenopausal women with metabolic syndrome was examined in a randomised clinical study. After six months of treatment, both brachial artery flow-mediated vasodilation (FMD) at age 50 and peak FMD were considerably higher in genistein receivers than in placebo recipients. In addition, genistein dramatically lowered the blood levels of total cholesterol, triglycerides, homocysteine, and visfatin, but blood adiponectin levels rose (Torres *et al.*, 2006). Another study was conducted at three different university medical facilities in Italy and used a randomised, double-blind, placebo-controlled design. At one year, fasting glucose, fasting insulin, and Homeostatic Model Assessment for Insulin Resistance (HOMA-IR) reduced in those who were given genistein, whereas they remained the same in those who were given placebo. Blood levels of total cholesterol, LDL-C, triglycerides, visfatin, and homocysteine were all found to be reduced when genistein was administered. However, high density lipoprotein cholesterol and adiponectin both saw statistically significant increases. In patients who received genistein, both systolic and diastolic blood pressure dropped significantly (Squadrito *et al.*, 2000). In postmenopausal individuals who had metabolic syndrome, ingestion of soy nuts for a shorter period of time resulted in decreased levels of inflammatory markers and an increase in plasma levels of nitric oxide (Azadbakht *et al.*, 2007). A randomised, parallel and single-centre 12-week nutritional intervention studies were also conducted in this investigation. In place of animal protein, whole soy meals comprising 30 g/day of soy protein were substituted. The incorporation of whole soy meals (30 g of protein per day) into a lipid-lowering diet dramatically reduced a set of biomarkers linked with cardiovascular risk (Ruscica *et al.*, 2018).

A clinical study reported that isoflavone consumption is related with enhanced vascular endothelial function and decreased carotid atherosclerotic load in people at high risk of cardiovascular events (Chan *et al.*, 2007). Atherosclerosis can be prevented by taking genistein, which has been shown to be effective. Zhang *et al.* (2017) also demonstrated that genistein might cure ox-LDL-induced inflammation in HUVECs by inhibiting the NF-κB signalling cascade via the miR-155/SOCS1 pathway (Zhang *et al.*, 2017a). Genistein reduced ROS and MDA formation by inhibiting SOD, CAT, glutathione, and glutathione peroxidase. The effects of genistein on sirtuin-1 overexpression were connected to the suppression of miR-34a, and their reversal required either a sirtuin-1 siRNA or a miR-34a mimic. It was also shown that manganese superoxide dismutase (MnSOD) and CAT's expressions were increased in the presence of genistein, which was linked to miR-34a/sirtuin-1-mediated nuclear translocation and deacetylation, as well as the higher expression of MnSOD. MiR-34a/sirtuin-1 and FoxO3a may play a significant role in HUVECs' ability to reverse the damage caused by LDL-induced oxidative stress (Zhang *et al.*, 2017b). Their recent study reported that genistein's protective effects on endothelium senescence via SIRT1 activation are likely related to reducing p66shc-mediated oxidative stress. In addition to inhibiting mitochondrial ROS formation, inactivated p66shc may cause Foxo3a accumulation in the nucleus, increasing the expression of target genes (such as MnSOD and CAT) that remove cellular ROS (Zhang *et al.*, 2022).

Genistein inhibits the endothelial cells-monocyte hyperglycemia-induced interaction, which is linked with lower endothelial production of chemokines and adhesion molecules, when used at physiologically appropriate dosages. Genistein's vasculoprotective properties are frequently attributed to its role in the cAMP/PKA signalling pathway, however this effect is not mediated by the ER-mediated signalling machinery. Genistein appears to be a promising new treatment for diabetes-induced inflammation and vascular dysfunction (Babu *et al.*, 2012).

Genistein's potential to protect endothelial cells via antioxidative effects stems from the fact that it reduces NADPH oxidase expression and activity. Western blotting of aortic endothelial cells from stroke-prone spontaneously hypertensive rats showed a concentration- and time-dependent reduction in p22phox NADPH oxidase subunit and angiotensin II type 1 (AT1) receptor expression. Genistein decreased superoxide generated by Ang II stimulation, inhibited nitrotyrosine synthesis, and lowered endothelin-1 production in an ELISA (Xu *et al.*, 2004).

3.3.2. *Soy isoflavones, in vitro and animal studies in cardiovascular diseases*

3.3.2.1. Adipocytes

Aziz *et al.* (2017) used genistein to find that genes involved in the differentiation of white (ACC) and brown/beige (Fasn, Fabp4, HSL, chemerin, and resistin) adipocytes were downregulated in 3T3-L1 adipocytes (CD-137 and UCP1). The transcript levels of genes involved in adipocyte differentiation (Cebp, Pgc1, and Sirt1) peaked at distinct times after genistein therapy. These responses were unaffected by the ER antagonist fulvestrant, suggesting that genistein does not work through the usual ER pathway. The Sirt1 inhibitor Ex-527 inhibited genistein's ability to increase mRNA expression of Uncoupling protein 1 (UCP1) and CCAAT-enhancer-binding proteins (Cebp), indicating that Sirt1 mediated this effect. Enhanced baseline oxygen consumption and the percentage contribution of proton leak to peak respiratory capacity were indicators of increased mitochondrial uncoupling in genistein-treated cells. (Aziz *et al.*, 2017). Additionally, Grossini *et al.* (2018) found that genistein increases cell survival, mitochondrial membrane potential, and differentiation and browning of human visceral pre-adipocytes. Modulation of Mitofusin-2 (Mfn2) expression in pre-adipocytes and adipocytes shows that genistein prevents ROS release and maintains mitochondrial function. (Grossini *et al.*, 2018).

Obesity-induced adipose inflammation is characterised by aberrant production of pro-inflammatory cytokines in white adipose tissues (Weisberg *et al.*, 2003). When differentiated 3T3-L1 adipocytes are co-cultured with RAW264 macrophages *in vitro*, pro-inflammatory cytokine genes and proteins such as CCL2, IL-6, and TNFα are upregulated (Suganami *et al.*, 2005; 2007). Sakamoto *et al.* (2016) found that daidzein activates PPARα and PPARγ and inhibits the JNK pathway, causing pro-inflammatory gene expression alterations in adipocyte and macrophage co-cultures. These outcomes may be helpful in reducing adipose inflammation, suggesting that daidzein administration might be an effective therapeutic method for chronic inflammation in obese adipose tissue (Sakamoto *et al.*, 2016).

3.3.2.2. Endothelial cells

Endothelial NO synthase (eNOS) synthesises NO from L-arginine and molecular oxygen to maintain vascular tone. eNOS activity and/or expression have been linked to stroke, hypertension, and atherosclerosis. It was shown that genistein works directly on vascular endothelial cells (EC) to improve eNOS activity and expression, which ultimately results in an increase in NO production (Liu *et al.*, 2004; Si and Liu, 2008; Räthel *et al.*, 2005). Si and Liu (2008) were the first to show that genistein upregulates eNOS expression and NO generation in human vascular endothelial cells via the cAMP/PKA/CREB/CRE pathway. In light of these findings, additional evaluation of this food-derived small molecule's potential in preventing or treating cardiovascular disease is now possible (Si and Liu, 2008). In addition, EVC-304 cells (a spontaneously-transformed line derived from a Japanese human umbilical vein endothelial cells culture) were protected against the cytotoxicity and apoptosis that were caused by hydrogen peroxide by the presence of soy isoflavones. Isoflavones derived from soy have the potential to have anti-apoptotic actions through the inhibition of reactive oxygen species (ROS) production, modulation of nuclear factor kappa B (NF-kB), and activation of a mitochondria-mediated apoptosis signalling pathway (Jin *et al.*, 2015). In ox-LDL-induced HUVECs, genistein was able to suppress the expression of NOX4 and reduce NADPH oxidase-dependent generation of superoxide. The addition of genistein to ox-LDL-induced HUVECs led to an increase in the radioactivity of BH4/BH2 as well as the expressions of GTP cyclohydrolase 1 (GCH1) and dihydrofolate reductase (DHFR). The action of genistein on ox-LDL-induced HUVECs was related with the sirtuin-1 pathway, which had the effect of attenuating eNOS uncoupling (Zhang *et al.*, 2016).

Atherosclerosis can be prevented with the use of genistein. Genistein was applied to HUVECs for six hours prior to exposure to ox-LDL for additional 24 hours in this investigation. As a result of this, genistein was able to reduce the levels of the oxidised LDL receptor, E-selectin and P-selectin, as well as monocyte chemotactic protein-1, interleukin 8 and vascular and intercellular adhesion molecules-1. Further studies showed that genistein's anti-inflammatory activity was linked to a reduction in miR-155 and an increase in SOCS1, and miR-155 mimics or SOCS1 siRNA performed similarly. Genistein also inhibited the NFκB signalling pathway, which contributed to its anti-cancer effects in the study. Using HUVECs, this study found that genistein reversed ox-LDL-induced inflammation by suppressing the NFκB signalling pathway via miR-155/SOCS1 suppression (Zhang *et al.*, 2017).

3.3.2.3. Vascular smooth muscle cells

Vascular smooth muscle cell (VSMC) cultures have also been studied in relation to isoflavones. After endothelial dysfunction, VSMC proliferation and migration are thought to be an important step in the development of atherosclerotic plaques. Soy isoflavone genistein boosts antioxidant enzyme activity and inhibits tyrosine kinase. Autophagy is a lysosomal-dependent mechanism for organelle and protein turnover. Autophagy prolongs survival under starvation. Genistein activated LKB1–AMPK signalling in VSMCs. Genistein enhanced LKB1 and AMPK phosphorylation in VSMCs. LKB1 and AMPK downregulated mTOR in VSMCs, inducing autophagy. Genistein-induced autophagy was reduced in dominant-negative AMPK-transfected cells but accelerated in constitutively active cells. Increasing perinuclear LC3-II puncta in genistein-treated VSMCs revealed enhanced autophagosome activity. Genistein-induced autophagy decreased adriamycin-induced SA-b-gal staining. These findings imply genistein-dependent autophagy reduces VSMC senescence via a LKB1–AMPK-dependent pathway (Lee *et al.*, 2016).

3.3.2.4. Animal studies

Menopause marks the beginning of a decline in women's previously enhanced resistance to vascular diseases. Vascular diseases including atherosclerosis and abdominal aortic aneurysm are more common in postmenopausal women. Vascular disease is a serious threat to postmenopausal women, yet there is currently no effective treatment to prevent it. Isoflavones are a possible useful dietary ingredient since they have a chemical structure similar to estrogen. Miyamoto and colleagues investigated the effect of isoflavones on aortic wall degeneration in ovariectomised (OVX) mice. However, isoflavones helped reduce the OVX group's increased loss of elastic fibers in the thoracic and abdominal aortas. There were more areas positive for MMP-2 and MMP-9 in the OVX group than in the control group. The presence of MMP-2 and MMP-9 positive areas were decreased by isoflavones compared to the OVX group. These results lend credence to the idea that isoflavones could inhibit MMP-2 and MMP-9 expression, hence lowering OVX-induced aortic wall degeneration (Miyamoto *et al.*, 2022).

Although there is evidence that soy isoflavones have positive health benefits, their ability to improve symptoms of the metabolic syndrome (MetS) has not been well investigated. Isoflavones' impact on lipid metabolism, inflammation, and oxidative stress has been studied in both *in vitro* and *in vivo* settings. The gene expression of cholesterol and lipid metabolism genes such as Scd-1 (27.7-fold), Cyp4a14 (35.2-fold), and Cyp4a10 (9.5-fold) was upregulated in Novasoy (NS)-fed nude mice, whereas the expression of anti-inflammatory genes such as Cebpd was downregulated (16.4-fold). C57BL/6J mice fed a high-fat (HF) diet supplemented with 0.4% (w/w) NS for 10 weeks had substantially lower percent weight growth and hepatic lipid accumulation compared to HF alone. In addition to reducing inflammatory cytokines, NS boosted lipid oxidation and antioxidant gene expression. Analysis of genistein's effects on oleic acid-induced lipid accumulation in HepG2 cells showed a dose-dependent inhibition of this process. The anti-inflammatory, antioxidant, and hypolipidemic effects of soy isoflavones may help relieve some of the symptoms of metabolic syndrome (Luo *et al.*, 2016). Both genistein and daidzein have been also shown to change LXR signalling both *in vivo* and *in vitro*, as well as reduce food intake and body weight increase in mice (Luo *et al.*, 2018).

There is a correlation between atherosclerosis susceptibility and food and other environmental variables. Soy protein-rich diets have been proposed as a preventative measure against atherosclerosis. Soy proteins or isoflavones, such genistein, may have an effect on health by decreasing cholesterol, preventing lipoprotein oxidation, or slowing cell proliferation. In C57BL/6 mice, ingestion of isoflavones from soy would lower plasma cholesterol and atherosclerosis, and in LDLR-null mice, it would reduce the amount of LDL that is oxidised (Kirk *et al.*, 1998). The fermented soybean product known as semen sojae preparatum (SSP) contains isoflavones as its primary active component. Through *in vivo* and *in vitro* tests, Guo *et al.* (2022) attempted to establish the efficiency of isoflavones from semen sojae preparatum (ISSP) in decreasing oxidative stress and its essential molecular pathways. ApoE/ mice were utilised to create atherosclerosis models by feeding them a high-fat diet, and endothelial cells were employed to create oxidative stress damage models by inducing ox-LDL. It was shown that ISSP's ability to repair endothelial cell oxidative damage and activate the Nrf2 signalling pathway was attenuated when the estrogen receptors GPR30 and ER were antagonised. The protective effect of ISSP against oxidative damage to endothelial cells was diminished after Nrf2 was downregulated through transient silencing with Nrf2-siRNA. This research suggests that ISSP's potential to protect against oxidative stress and atherosclerosis may involve the GPR30 and ER estrogen receptors and occur via the Nrf2 signalling pathway (Guo *et al.*, 2022).

Analysing epidemiological, clinical, and experimental evidence, we found that isoflavones (in the human diet improves cardiovascular health in several ways, including: (i) reduced triglyceride, cholesterol, low-density lipoprotein, high-density lipoprotein, and apolipoprotein B levels in postmenopausal women's blood; (ii) reduced inflammation markers and improved endothelial function; (iii) reduced systolic and diastolic blood pressure in genistein patients; (iv) reduced carotid atherosclerotic load in those at high risk for cardiovascular events.

To summarise, isoflavones, which include genistein and daidzein, have been shown to have protective benefits against metabolic syndrome through their actions on adipocytes, endothelial cells, and vascular smooth muscle cells. **In adipocytes**, genistein promotes survival, mitochondrial membrane potential, differentiation, and browning. Genistein improves visceral pre-adipocyte survival, membrane potential, differentiation, and browning. Daidzein activates PPARs and inhibits JNK, altering pro-inflammatory gene expression in adipocyte and macrophage co-cultures. **In endothelial cells**, genistein improves eNOS activity and expression, increasing NO generation. Soy isoflavones may decrease ROS generation, modulate NF-kB, and activate a mitochondria-mediated apoptosis signalling pathway. Genistein inhibited eNOS uncoupling in ox-LDL-induced HUVECs via the sirtuin-1 pathway. Genistein reduced oxidised LDL receptor, E-selectin, P-selectin, monocyte chemotactic protein-1, interleukin 8, and vascular and intercellular adhesion molecules-1. In HUVECs, genistein reduced ox-LDL-induced inflammation by decreasing NFκB signalling via miR-155/SOCS1. **In vascular smooth muscle cells**, genistein inhibits VSMC senescence by activating autophagy via a LKB1-AMPK-dependent mechanism.

3.4. *Role of bioactive compounds in the prevention of cancers*

Cancer is a devastating disease whose burden is measured in terms of incidence and mortality. It is a condition whereby the body cells multiply in an abnormal and uncontrollable manner. Approximately 11 million individuals around the world are diagnosed with this disease annually and 6.7 million of them succumb to this illness (George *et al.*, 2021). There are more than 200 types of cancer with the major types being breast, colorectal, and lung cancers. The ones with the highest mortality rates are stomach, liver, and lung cancers. Breast cancer is the most prevalent form of cancer in females accounting for almost 3 million deaths in women and 13.9% cancer-related deaths (Mustafa *et al.*, 2017). In males, prostate cancer is the most prevalent.

The signs and symptoms vary based on size, type, location, and stage of the disease. Diagnosis can be determined by biopsy results and treatment options depending on the type and severity of the tumour. Owing to the challenges associated with modern cancer treatment such as increased mortality rate, disease severity, and loss of patient quality of life, there is a need to develop a new and alternative treatment and prevention approach for patients. Death rates are higher in developing countries due to late diagnosis and lack of public awareness (George *et al.*, 2021). Functional foods and bioactive compounds are important in the treatment and prevention of cancer. Experts recommend taking healthy diet as a way of preventing the development of this illness. The following section will describe the role of different bioactive compounds (carotenoids such as α-carotene, β-carotene, lycopene, lutein, and zeaxanthin) in prevention of cancers, particularly breast cancer and prostate cancer.

3.4.1. *Breast cancer and carotenoids*

3.4.1.1. Clinical studies

Over 685,000 people died from breast cancer in 2020, with 2.3 million being diagnosed. As of the end of the year 2020, 7.8 million women have been diagnosed with breast cancer within the previous five years, making it the most frequent disease in the world (WHO, 2020 https://www.who.int/news-room/fact-sheets/detail/breast-cancer). Fruits and vegetables' carotenoid antioxidants are one of the few modifiable risk factors for primary cancer prevention at the individual level. According to the findings of a meta-analysis and meta-regression study, significant dose-response correlations were found in the increased consumption of dietary and total β-carotene with a lower breast cancer risk in both cohort and case-control studies. Women who ingest more β-carotene through their diets may have a lower risk of developing breast cancer (Hu *et al.*, 2012). Plasma carotenoids were shown to have suggestive inverse relationships with the risk of premalignant breast disease in younger women, which is consistent with the inverse connections documented for invasive breast cancer. It is possible that carotenoids have a role in the early stages of breast cancer (Cohen *et al.*, 2017). Some studies showed that postmenopausal women who have high levels of plasma α-carotene appear to have a lower chance of developing breast cancer (Wang *et al.*, 2015; Epplein *et al.*, 2009; Kabat *et al.*, 2009). In a meta-analysis, there was not found to be any relationship between dietary consumption of carotenoids and the risk of breast cancer, with the exception of a marginal reduction in risk associated with dietary β-carotene. Blood levels of total carotenoids, β-carotene, α-carotene, and lutein were shown to be more significantly related with reduced breast cancer risk (Aune *et al.*, 2012). A recent nested case-control study in the Nurse' Health Studies demonstrated that high-risk women were more likely to have inverse correlations between circulating carotenoids and breast cancer risk, as determined by their germline genetic composition or mammographic density (Peng *et al.*, 2021). A link between carotenoids and breast cancer risk was found in a recent systematic study by Peraita-Costa and coworkers (2022). Carotenoids may help prevent breast cancer if they are included in one's diet. (Peraita-Costa *et al.*, 2022). A growing body of large-scale prospective cohorts study has connected dietary and circulatory carotenoids to a lower risk of breast cancer. A comprehensive pooled study of over 3000 case participants from eight prospective studies demonstrated statistically significant inverse relationships between circulating levels of individual and total carotenoids and breast cancer risk. Specifically, it was shown that β-carotene, α-carotene, lutein + zeaxanthin, and lycopene all had negative relationships, whereas beta-cryptoxanthin does not. (Eliassen *et al.*, 2012). Zhang *et al.* (2012) showed that intakes of β-carotene, α-carotene, and lutein/zeaxanthin were inversely linked with risk of estrogen receptor (ER)-negative breast cancer, but this association was not seen with ER+ breast cancer (Zhang *et al.*, 2012). Another study also reported that greater plasma concentrations of both β-carotene and α-carotene are related to a reduced chance of developing ER- breast cancer (Bakker *et al.*, 2016). In a large prospective study was carried out by Assar EA *et al.* (2015) over a period of 20 years of follow-up, the researchers found that women who had high plasma carotenoids had a lower risk of developing breast cancer, in particular the more aggressive and ultimately fatal forms of the disease (Eliassen *et al.*, 2015).

3.4.1.2. *In vitro* studies

Carotenoids suppress cell development and differentiation, activate a protein kinase cascade, and are anticarcinogenic. In a study published in 2018, Gong *et al.* showed that lutein suppressed breast cancer cell lines but not normal mammary epithelial cells. Furthermore, β-carotene and astaxanthin, two other carotenoids having biological activity in other systems, had no effect. Lutein causes cell cycle arrest and death in TNBC cells. Unlike basic mammary epithelial cells, lutein only raised intracellular ROS levels in TNBC cells. This enhanced ROS generation which appears to mediate the suppression of lutein-mediated proliferation in breast cancer cells. Additional findings showed that lutein therapy activates the p53 signalling pathway and raises HSP60 levels, both of which may lead to a reduction in TNBC cell proliferation (Gong *et al.*, 2018). It was also suggested that the antiproliferative action of lutein was mediated via activation of the Nuclear factor-erythroid factor 2-related factor 2/antioxidant responsive element (NrF2/ARE) pathway and inhibition of the NFκB signalling pathway (Chang *et al.*, 2018). In addition, lutein suppressed hypoxia-induced breast cancer cell proliferation, invasion, and migration via decreasing expression of hairy and enhancer of split-1 (HES1). These results suggest that lutein may one day be used to develop novel therapeutic approaches for the treatment of breast cancer, as it seems to control HES1 through the Hypoxia-Inducible Factor (HIF)-1 and NOTCH signalling molecular pathways (Li *et al.*, 2018).

Another carotenoid β-carotene triggered apoptosis in a dose-dependent manner in MCF-7, MDA-MB-235 and MDA-MB-231 cells. After 48 hours, β-carotene increased apoptosis more rapidly than lycopene (Gloria *et al.*, 2014). Purified β-carotene decreased MCF-7 cell viability in a dose-dependent manner that was highly linked with alterations in cell shape.

Apoptotic cells were found to be more prevalent in cells treated with β-carotene (1 μM). This activation of apoptosis was linked to enhanced caspase-3 activity. Protein expression tests revealed that carotene at a concentration of 1 μM efficiently reduces the expression of the anti-apoptotic proteins Bcl-2 and PARP, as well as the survival protein NFκB. Akt and ERK1/2, intracellular growth signalling proteins, were also blocked. β-carotene inhibits Akt activation, reducing bad phosphorylation. It also lowered SOD-2, Nrf-2, and ER stress marker XBP-1 protein levels. These findings show that β-carotene plays an important role in MCF-7 cells even at low physiological concentrations, which explains its potent anti-cancer effect (Sowmya Shree *et al.*, 2017). Another study by Peng *et al.* (2017) demonstrated that lycopene suppresses MCF-7 cell growth and promoted apoptosis *in vitro*, potentially via modulating the expression of p53 and Bax (Peng *et al.*, 2017).

Collective speaking, these findings imply interesting prospects to integrate recommendations for high-carotenoids diets into breast cancer prevention for women who are at greater risk, however, given the majority of circulating carotenoids derive from dietary sources.

3.4.2. *Prostate cancer and carotenoids (lycopene and α-carotene)*

3.4.2.1. Clinical studies

It is anticipated that in 2020, there will be 1,414,000 new instances of prostate cancer, with a death toll of 375,304, making it the second most often diagnosed disease in males globally. Prostate cancer is the most common form of cancer in 112 countries and the major cause of cancer death in 48 (Sung *et al.*, 2021). It is important to remember that as the population ages and the economy grows, the prevalence of prostate cancer is expected to rise (Culp *et al.*, 2020). Diet, for instance, can have a significant effect in prostate cancer incidence (Bagheri *et al.*, 2018).

Epidemiological research has shown that men who consume a diet rich in lycopene have a lower risk of developing prostate cancer than those who do not (Giovannucci, 2005). Prostate cancer was shown to be associated with greater plasma lycopene concentrations in older men and those without a family history of the disease. It was shown that tomato products may be more effective in protecting against sporadic prostate cancer than those with a larger familial or genetic component. β-carotene consumption may also reduce the risk of prostate cancer in younger men, according to a nest control study by (Wu *et al.*, 2004). There is some evidence that males in Vietnam who eat more lycopene-rich foods like tomatoes and carrots have a decreased chance of developing prostate cancer (Van Hoang *et al.*, 2018).

Lycopene has been found as a possible anticancer drug with antioxidant characteristics. Researchers also looked at dietary consumption and blood levels of carotenoids in connection to prostate cancer (PCa) risk using a systematic review and dose-response meta-analysis (Chen *et al.*, 2015). With a focus on the dose-response curve shape, another systematic review and meta-analysis also indicated that higher lycopene consumption/circulating concentration is associated with a lower risk of prostate cancer (Giovannucci, 2005). Additionally, both lycopene and α-carotene, but not β-carotene, reduced the risk of prostate cancer. There was no evidence that either α-carotene or lycopene might reduce the incidence of advanced prostate cancer (Wang *et al.*, 2015). A pooled analysis of 15 studies conducted by Key *et al.* (2015) showed that α-tocopherol and lycopene were shown to be related with a lower risk of aggressive prostate cancer (Key *et al.*, 2015). In spite of this, neither the control nor any of the treatments (including those with tomato extracts or lycopene) significantly changed PSA levels. However, larger, more stringent clinical trials are needed to fully understand the influence of tomato extract or lycopene on PSA level (Sharifi-Zahabi *et al.*, 2022). The ProDiet randomised control trial reported that the serum metabolome of men at high risk for prostate cancer was changed after an intervention to enhance lycopene consumption. There is a correlation between lycopene consumption and a lower chance of developing prostate cancer, as shown by these findings. Lycopene inhibited pyruvate production, which led to a reduction in energy expenditure. (Beynon *et al.*, 2019).

Castration-resistant prostate cancer (CRPC) can be effectively treated with the medicine enzalutamide, however many patients will develop resistance to enzalutamide within a very short period of time. Recently, a study showed that lycopene's ability to increase the antitumour effects of enzalutamide in CRPC may be linked to its ability to decrease androgen receptor (AR) protein levels via lycopene-mediated inhibition of the AKT/EZH2 pathway (Chen *et al.*, 2022). Interestingly, a recent cross-sectional study examined the relationship between lycopene consumption from daily foods and the risk of prostate cancer. The outcomes of this study show that enough lycopene consumption may lessen the incidence of prostate cancer. However, this connection was only detected in Non-Hispanic White (NHW) individuals. This difference may have several causes. Consistent with a prior study indicating that Non-Hispanic Black (NHB) persons consume less lycopene than NHW individuals, they discovered that only 16.8% of NHB respondents had adequate lycopene consumption, compared to 29.4% of NHW respondents. Using predetermined independent PCa risk variables, stratification analysis indicated that living alone is a significant barrier for NHB persons to ingest lycopene. Future research should concentrate on the connection between living status and lycopene consumption for the prevention of PCa risk in NHB persons (Lu *et al.*, 2021). In addition, several clinical trials have demonstrated the potential advantages of lycopene supplementation, including reductions in tumour volume and prostate-specific antigen (PSA) levels. Supplementing with lycopene has shown promise in a number of clinical studies, with benefits including decreased tumour volume (Kucuk *et al.*, 2001; 2002) and PSA levels (Ansari and Gupta, 2003).

3.4.2.2. *In vitro* and animal studies

In vitro and *in vivo* studies have been done to find out what lycopene does. Lycopene has been shown to inhibit DNA synthesis. The inhibitory effects of lycopene in primary prostatic epithelial cell cultures and the findings of a pilot phase II clinical investigation on established prostate cancer showed a substantial and sustained effect on PSA over 1 year (Barber *et al.*, 2006). Additionally, lycopene has been proved to modulate cell cycle. The proliferation of normal human prostate epithelial cells (PrEC) was considerably suppressed by lycopene, and this effect was dosage dependent. The cells had stopped dividing and were in the G0/G1 phase, as determined by flow cytometry. The expression of cyclin D1 was suppressed, confirming this impact, although cyclin E levels were unaffected

(Obermüller-Jevic *et al.*, 2003). Isoflavones are promising chemopreventive agents with multiple cellular effects, both genomic and non-genomic, including, but not limited to, impacting insulin-like growth factor induced (IGF-I) (*Nahum et al.*, 2006), having an anti-inflammatory impact (Jiang *et al.*, 2018), bringing about cell death through apoptosis (Jiang *et al.*, 2018; Soares *et al.*, 2017) and enhancing gap junction protein (Aust *et al.*, 2003). An *in vitro* study by shows that lycopene, at a level that is theoretically feasible in humans, can affect the cell adhesion and migratory capabilities of cancer cells. Their findings add to the growing body of evidence supporting lycopene's potential to act as a cancer-preventative chemotherapy agent in the case of prostate cancer (Elgass *et al.*, 2014).

Researchers using rodent models of prostate cancer have studied the effects of lycopene on cancer development. When given to BALB/c nude mice orally for three weeks, lycopene dramatically reduced the development of DU145 tumour xenografts. This results was consistent with *in vitro* data on growth suppression in this cell line, suggesting that lycopene operates primarily on the phases of tumour development and progression in addition to its robust antioxidant effects (Elgass *et al.*, 2014). Another study demonstrated that lycopene was also discovered to reduce expression of IGI-I and interleukin 6 (IL-6) in the prostate (Siler *et al.*, 2004).

Several laboratories have looked at the effects of lycopene on prostate cancer in rodent models. They have found that androgens affect lycopene metabolism. For example, castrated rats store twice as much lycopene in their livers as intact rats, which stops testosterone from being activated locally (Boileau *et al.*, 2001). It was also found that lycopene stops insulin-like growth factor I and IL-6 from being made in prostate tissue (Siler *et al.*, 2004).

4. Conclusions and Perspectives

Several bioactive compounds in crops and their potential health benefits to humans have been reported and assessed. Different types of chronic illnesses and metabolic disorders can be protected against by their special antioxidant, anti-inflammatory, and anti-carcinogenic characteristics, as well as their related physiological and cellular effects. They are mostly found in plant foods like fruits and vegetables, and their inclusion in diets that have been shown to have positive health impacts makes them promising candidates for the creation of novel functional foods with, among other things, protective and preservation qualities. In order to obtain these bioactive compounds and reap the positive health advantages, consumers should eat more fruits and vegetables, but more study is needed to determine the precise mechanisms of their biological functions.

References

Ansari, M.S. and Gupta, N.P., 2003. A comparison of lycopene and orchidectomy vs orchidectomy alone in the management of advanced prostate cancer. *BJU Int.*, 92(4): 375–378; discussion 378. https://doi.org/10.1046/j.1464-410x.2003.04370.x

Appel, L.J., 2003. The effects of protein intake on blood pressure and cardiovascular disease. *Curr. Opin. Lipidol.*, 14(1): 55–59.

Aune, D., Chan, D.S.M., Vieira, A.R., Navarro Rosenblatt, D.A., Vieira, R., Greenwood, D.C., and Norat, T., 2012. Dietary compared with blood concentrations of carotenoids and breast cancer risk: A systematic review and meta-analysis of prospective studies. *Am. J. Clin. Nutr.*, 96(2): 356–373. https://doi.org/10.3945/ajcn.112.034165

Aust, O., Ale-Agha, N., Zhang, L., Wollersen, H., Sies, H., and Stahl, W., 2003. Lycopene oxidation product enhances gap junctional communication. *Food Chem. Toxicol.*, 41(10): 1399–1407. https://doi.org/10.1016/s0278-6915(03)00148-0

Azadbakht, L., Kimiagar, M., Mehrabi, Y., Esmaillzadeh, A., Hu, F.B., and Willett, W.C., 2007. Soy consumption, markers of inflammation, and endothelial function: A cross-over study in postmenopausal women with the metabolic syndrome. *Diabetes Care*, 30(4): 967–973. https://doi.org/10.2337/dc06-2126

Aziz, S.A., Wakeling, L.A., Miwa, S., Alberdi, G., Hesketh, J.E., and Ford, D., 2017. Metabolic programming of a beige adipocyte phenotype by genistein. *Mol. Nutr. Food Res.*, 61(2). https://doi.org/10.1002/mnfr.201600574

Bagheri, A., Nachvak, S.M., Rezaei, M., Moravridzade, M., Moradi, M., and Nelson, M., 2018. Dietary patterns and risk of prostate cancer: A factor analysis study in a sample of Iranian men. *Health Promot. Perspect.*, 8(2): 133–138. https://doi.org/10.15171/hpp.2018.17

Bakker, M.F., Peeters, P.H., Klaasen, V.M., Bueno-de-Mesquita, H.B., Jansen, E.H., Ros, M.M., Travier, N., Olsen, A., Tjønneland, A., Overvad, K., Rinaldi, S., Romieu, I., Brennan, P., Boutron-Ruault, M.-C., Perquier, F., Cadeau, C., Boeing, H., Aleksandrova, K., Kaaks, R. *et al.*, 2016. Plasma carotenoids, vitamin C, tocopherols, and retinol and the risk of breast cancer in the European Prospective Investigation into Cancer and Nutrition cohort. *Am. J. Clin. Nutr.*, 103(2): 454–464. https://doi.org/10.3945/ajcn.114.101659

Barber, N.J., Zhang, X., Zhu, G., Pramanik, R., Barber, J.A., Martin, F.L., Morris, J.D.H., and Muir, G.H., 2006. Lycopene inhibits DNA synthesis in primary prostate epithelial cells *in vitro* and its administration is associated with a reduced prostate-specific antigen velocity in a phase II clinical study. *Prostate Cancer Prostatic Dis.*, 9(4): 407–413. https://doi.org/10.1038/sj.pcan.4500895

Beynon, R.A., Richmond, R.C., Santos Ferreira, D.L., Ness, A.R., May, M., Smith, G.D., Vincent, E.E., Adams, C., Ala-Korpela, M., Würtz, P., Soidinsalo, S., Metcalfe, C., Donovan, J.L., Lane, A.J., Martin, R.M., ProtecT Study Group and PRACTICAL consortium, 2019. Investigating the effects of lycopene and green tea on the metabolome of men at risk of prostate cancer: The ProDiet randomised controlled trial. *Int. J. Cancer*, 144(8): 1918–1928. https://doi.org/10.1002/ijc.31929

Boileau, T.W., Clinton, S.K., Zaripheh, S., Monaco, M.H., Donovan, S.M., and Erdman, J.W., 2001. Testosterone and food restriction modulate hepatic lycopene isomer concentrations in male F344 rats. *J. Nutr.*, 131(6): 1746–1752. https://doi.org/10.1093/jn/131.6.1746

Chang, J., Zhang, Y., Li, Y., Lu, K., Shen, Y., Guo, Y., Qi, Q., Wang, M., and Zhang, S., 2018. NrF2/ARE and NF-κB pathway regulation may be the mechanism for lutein inhibition of human breast cancer cell. *Future Oncol. (London, England)*, 14(8): 719–726. https://doi.org/10.2217/fon-2017-0584

Chen, P., Zhang, W., Wang, X., Zhao, K., Negi, D.S., Zhuo, L., Qi, M., Wang, X., and Zhang, X., 2015. Lycopene and risk of prostate cancer: A systematic review and meta-analysis. *Medicine*, 94(33): e1260. https://doi.org/10.1097/MD.0000000000001260

Chiva-Blanch, G., Jiménez, C., Pinyol, M., Herreras, Z., Catalán, M., Martínez-Huélamo, M., Lamuela-Raventos, R.M., Sala-Vila, A., Cofán, M., Gilabert, R., Jiménez, A., and Ortega, E., 2020. 5-cis-, Trans- and total lycopene plasma concentrations inversely relate to atherosclerotic plaque burden in newly diagnosed type 2 diabetes subjects. *Nutrients*, 12(6): E1696. https://doi.org/10.3390/nu12061696

Cladis, D.P., Simpson, A.M.R., Cooper, K.J., Nakatsu, C.H., Ferruzzi, M.G., and Weaver, C.M., 2021. Blueberry polyphenols alter gut microbiota and phenolic metabolism in rats. *Food Funct.*, 12(6): 2442–2456. https://doi.org/10.1039/d0fo03457f

Cohen, K., Liu, Y., Luo, J., Appleton, C.M., and Colditz, G.A., 2017. Plasma carotenoids and the risk of premalignant breast disease in women aged 50 and younger: A nested case-control study. *Breast Cancer Res. Treat.*, 162(3): 571–580. https://doi.org/10.1007/s10549-017-4152-5

Cuevas-Sierra, A., Ramos-Lopez, O., Riezu-Boj, J.I., Milagro, F.I., and Martinez, J.A., 2019. Diet, gut microbiota, and obesity: links with host genetics and epigenetics and potential applications. *Adv. Nutr.*, 10(Suppl 1): S17–S30. https://doi.org/10.1093/advances/nmy078

Culp, M.B., Soerjomataram, I., Efstathiou, J.A., Bray, F., and Jemal, A., 2020. Recent global patterns in prostate cancer incidence and mortality rates. *Eur. Urol.*, 77(1): 38–52. https://doi.org/10.1016/j.eururo.2019.08.005

Curtis, P.J., Berends, L., van der Velpen, V., Jennings, A., Haag, L., Chandra, P., Kay, C.D., Rimm, E.B., and Cassidy, A., 2022. Blueberry anthocyanin intake attenuates the postprandial cardiometabolic effect of an energy-dense food challenge: Results from a double blind, randomized controlled trial in metabolic syndrome participants. *Clin. Nutr.*, 41(1): 165–176. https://doi.org/10.1016/j.clnu.2021.11.030

Curtis, P.J., van der Velpen, V., Berends, L., Jennings, A., Feelisch, M., Umpleby, A.M., Evans, M., Fernandez, B.O., Meiss, M.S., Minnion, M., Potter, J., Minihane, A.-M., Kay, C.D., Rimm, E.B., and Cassidy, A., 2019. Blueberries improve biomarkers of cardiometabolic function in participants with metabolic syndrome-results from a 6-month, double-blind, randomized controlled trial. *Am. J. Clin. Nutr.*, 109(6): 1535–1545. https://doi.org/10.1093/ajcn/nqy380

Dosi, R., Bhatt, N., Shah, P., and Patell, R., 2014. Cardiovascular disease and menopause. *J. Clin. Diagn. Res.*, 8(2): 62–64. https://doi.org/10.7860/JCDR/2014/6457.4009

Duan, L., Liang, C., Li, X., Huang, Z., Liu, S., Wu, N., and Jia, D., 2019. Lycopene restores the effect of ischemic postconditioning on myocardial ischemia-reperfusion injury in hypercholesterolemic rats. *Int. J. Mol. Med.*, 43(6): 2451–2461. https://doi.org/10.3892/ijmm.2019.4166

Elgass, S., Cooper, A., and Chopra, M., 2014. Lycopene treatment of prostate cancer cell lines inhibits adhesion and migration properties of the cells. *Int. J. Med. Sci.*, 11(9): 948–954. https://doi.org/10.7150/ijms.9137

Eliassen, A.H., Hendrickson, S.J., Brinton, L.A., Buring, J.E., Campos, H., Dai, Q., Dorgan, J.F., Franke, A.A., Gao, Y., Goodman, M.T., Hallmans, G., Helzlsouer, K.J., Hoffman-Bolton, J., Hultén, K., Sesso, H.D., Sowell, A.L., Tamimi, R.M., Toniolo, P., Wilkens, L.R. *et al.*, 2012. Circulating carotenoids and risk of breast cancer: pooled analysis of eight prospective studies. *J. Natl. Cancer Inst.*, 104(24): 1905–1916. https://doi.org/10.1093/jnci/djs461

Eliassen, A.H., Liao, X., Rosner, B., Tamimi, R.M., Tworoger, S.S., and Hankinson, S.E., 2015. Plasma carotenoids and risk of breast cancer over 20 y of follow-up. *Am. J. Clin. Nutr.*, 101(6): 1197–1205. https://doi.org/10.3945/ajcn.114.105080

Engelhard, Y.N., Gazer, B., and Paran, E., 2006. Natural antioxidants from tomato extract reduce blood pressure in patients with grade-1 hypertension: A double-blind, placebo-controlled pilot study. *Am. Heart J.*, 151(1): 100. https://doi.org/10.1016/j.ahj.2005.05.008

Epplein, M., Shvetsov, Y.B., Wilkens, L.R., Franke, A.A., Cooney, R.V., Le Marchand, L., Henderson, B.E., Kolonel, L.N., and Goodman, M.T., 2009. Plasma carotenoids, retinol, and tocopherols and postmenopausal breast cancer risk in the Multiethnic Cohort Study: A nested case-control study. *Breast Cancer Res.*, 11(4): R49. https://doi.org/10.1186/bcr2338

Esposito, K., Ciotola, M., and Giugliano, D., 2005. Inflammation warms up the metabolic syndrome. *Arterioscler. Thromb. Vasc. Biol.*, 25(11): e143. https://doi.org/10.1161/01.ATV.0000185830.17528.e3

Esposito, K. and Giugliano, D., 2004. The metabolic syndrome and inflammation: association or causation? *Nutr. Metab. Cardiovasc. Dis.*, 14(5): 228–232. https://doi.org/10.1016/s0939-4753(04)80048-6

Fan, S., Sun, J.-B., Li, R., Song, X., and Li, J., 2019. Lycopene protects myocardial ischemia injury through anti-apoptosis and anti-oxidative stress. *Eur. Rev. Med. Pharmacol. Sci.*, 23(7): 3096–3104. https://doi.org/10.26355/eurrev_201904_17593

Gajendragadkar, P.R., Hubsch, A., Mäki-Petäjä, K.M., Serg, M., Wilkinson, I.B., and Cheriyan, J., 2014. Effects of oral lycopene supplementation on vascular function in patients with cardiovascular disease and healthy volunteers: A randomised controlled trial. *PloS One*, 9(6): e99070. https://doi.org/10.1371/journal.pone.0099070

George, B.P., Chandran, R., and Abrahamse, H., 2021. Role of phytochemicals in cancer chemoprevention: Insights. *Antioxidants*, 10(9): 1455. https://doi.org/10.3390/antiox10091455

Gerhard, M., Roddy, M.A., Creager, S.J., and Creager, M.A., 1996. Aging progressively impairs endothelium-dependent vasodilation in forearm resistance vessels of humans. *Hypertension (Dallas, Tex.: 1979)*, 27(4): 849–853. https://doi.org/10.1161/01.hyp.27.4.849

Giovannucci, E., 2005. Tomato products, lycopene, and prostate cancer: A review of the epidemiological literature. *J. Nutr.*, 135(8): 2030S–2031S. https://doi.org/10.1093/jn/135.8.2030S

Gloria, N.F., Soares, N., Brand, C., Oliveira, F.L., Borojevic, R., and Teodoro, A.J., 2014. Lycopene and beta-carotene induce cell-cycle arrest and apoptosis in human breast cancer cell lines. *Anticancer Res.*, 34(3): 1377–1386.

Gong, X., Smith, J.R., Swanson, H.M., and Rubin, L.P., 2018. Carotenoid lutein selectively inhibits breast cancer cell growth and potentiates the effect of chemotherapeutic agents through ROS-mediated mechanisms. *Molecules (Basel, Switzerland)*, 23(4): E905. https://doi.org/10.3390/molecules23040905

Grandl, G. and Wolfrum, C., 2018. Hemostasis, endothelial stress, inflammation, and the metabolic syndrome. *Seminars in Immunopathology*, 40(2): 215–224. https://doi.org/10.1007/s00281-017-0666-5

Grossini, E., Farruggio, S., Raina, G., Mary, D., Deiro, G., and Gentilli, S., 2018. Effects of genistein on differentiation and viability of human visceral adipocytes. *Nutrients*, 10(8): E978. https://doi.org/10.3390/nu10080978

Guo, J., Ma, J., Cai, K., Chen, H., Xie, K., Xu, B., Quan, D., and Du, J., 2022. Isoflavones from Semen sojae preparatum improve atherosclerosis and oxidative stress by modulating Nrf2 signaling pathway through estrogen-like effects. *Evid. Based Complement. Alternat. Med.*, 2022, 4242099. https://doi.org/10.1155/2022/4242099

Han, G.-M. and Liu, P., 2017. Higher serum lycopene is associated with reduced prevalence of hypertension in overweight or obese adults. *Eur. J. Integr. Med.*, 13, 34–40. https://doi.org/10.1016/j.eujim.2017.07.002

Hu, F., Wang Yi, B., Zhang, W., Liang, J., Lin, C., Li, D., Wang, F., Pang, D., and Zhao, Y., 2012. Carotenoids and breast cancer risk: A meta-analysis and meta-regression. *Breast Cancer Res. Treat.*, 131(1): 239–253. https://doi.org/10.1007/s10549-011-1723-8

Huang, Y., Cao, S., Nagamani, M., Anderson, K.E., Grady, J.J., and Lu, L.-J.W., 2005. Decreased circulating levels of tumor necrosis factor-alpha in postmenopausal women during consumption of soy-containing isoflavones. *J. Clin. Endocrinol. Metabol.*, 90(7): 3956–3962. https://doi.org/10.1210/jc.2005-0161

Irace, C., Marini, H., Bitto, A., Altavilla, D., Polito, F., Adamo, E.B., Arcoraci, V., Minutoli, L., Di Benedetto, A., Di Vieste, G., de Gregorio, C., Gnasso, A., Corrao, S., Licata, G., and Squadrito, F., 2013. Genistein and endothelial function in postmenopausal women with metabolic syndrome. *Eur. J. Clin. Invest.*, 43(10): 1025–1031. https://doi.org/10.1111/eci.12139

Jiang, L.-N., Liu, Y.-B., and Li, B.-H., 2018. Lycopene exerts anti-inflammatory effect to inhibit prostate cancer progression. *Asian J. Androl.* https://doi.org/10.4103/aja.aja_70_18

Jin, L., Zhao, X., Qin, Y., Zhu, W., Zhang, W., Liu, A., and Luo, Z., 2015. Soy isoflavones protect against H_2O_2-induced injury in human umbilical vein endothelial cells. *Mol. Med. Rep.*, 12(3): 4476–4482. https://doi.org/10.3892/mmr.2015.3949

Johnson, S.A., Feresin, R.G., Navaei, N., Figueroa, A., Elam, M.L., Akhavan, N.S., Hooshmand, S., Pourafshar, S., Payton, M.E., and Arjmandi, B.H., 2017. Effects of daily blueberry consumption on circulating biomarkers of oxidative stress, inflammation, and antioxidant defense in postmenopausal women with pre- and stage 1-hypertension: A randomized controlled trial. *Food Funct.*, 8(1): 372–380. https://doi.org/10.1039/c6fo01216g

Johnson, S.A., Figueroa, A., Navaei, N., Wong, A., Kalfon, R., Ormsbee, L.T., Feresin, R.G., Elam, M.L., Hooshmand, S., Payton, M.E., and Arjmandi, B.H., 2015. Daily blueberry consumption improves blood pressure and arterial stiffness in postmenopausal women with pre- and stage 1-hypertension: A randomized, double-blind, placebo-controlled clinical trial. *J. Acad. Nutr. Diet.*, 115(3): 369–377. https://doi.org/10.1016/j.jand.2014.11.001

Kabagambe, E.K., Furtado, J., Baylin, A., and Campos, H., 2005. Some dietary and adipose tissue carotenoids are associated with the risk of nonfatal acute myocardial infarction in Costa Rica. *J. Nutr.*, 135(7): 1763–1769. https://doi.org/10.1093/jn/135.7.1763

Kabat, G.C., Kim, M., Adams-Campbell, L.L., Caan, B.J., Chlebowski, R.T., Neuhouser, M.L., Shikany, J.M., Rohan, T.E., and WHI Investigators, 2009. Longitudinal study of serum carotenoid, retinol, and tocopherol concentrations in relation to breast cancer risk among postmenopausal women. *Am. J. Clin. Nutr.*, 90(1): 162–169. https://doi.org/10.3945/ajcn.2009.27568

Kalea, A.Z., Clark, K., Schuschke, D.A., and Klimis-Zacas, D.J., 2009. Vascular reactivity is affected by dietary consumption of wild blueberries in the Sprague-Dawley rat. *J. Med. Food*, 12(1): 21–28. https://doi.org/10.1089/jmf.2008.0078

Kardinaal, A.F., Kok, F.J., Ringstad, J., Gomez-Aracena, J., Mazaev, V.P., Kohlmeier, L., Martin, B.C., Aro, A., Kark, J.D., and Delgado-Rodriguez, M., 1993. Antioxidants in adipose tissue and risk of myocardial infarction: The EURAMIC Study. *Lancet (London, England)*, 342(8884): 1379–1384. https://doi.org/10.1016/0140-6736(93)92751-e

Karppi, J., Laukkanen, J.A., Mäkikallio, T.H., and Kurl, S., 2012. Low serum lycopene and β-carotene increase risk of acute myocardial infarction in men. *Eur. J. Public Health*, 22(6): 835–840. https://doi.org/10.1093/eurpub/ckr174

Key, T.J., Appleby, P.N., Travis, R.C., Albanes, D., Alberg, A.J., Barricarte, A., Black, A., Boeing, H., Bueno-de-Mesquita, H.B., Chan, J.M., Chen, C., Cook, M.B., Donovan, J.L., Galan, P., Gilbert, R., Giles, G.G., Giovannucci, E., Goodman, G.E., Goodman, P.J. *et al.*, 2015. Endogenous hormones nutritional biomarkers prostate cancer collaborative group. Carotenoids, retinol, tocopherols, and prostate cancer risk: Pooled analysis of 15 studies. *Am. J. Clin. Nutr.*, 102(5): 1142–1157. https://doi.org/10.3945/ajcn.115.114306

Kirk, E.A., Sutherland, P., Wang, S.A., Chait, A., and LeBoeuf, R.C., 1998. Dietary isoflavones reduce plasma cholesterol and atherosclerosis in C57BL/6 mice but not LDL receptor–deficient mice. *J. Nutr.*, 128(6): 954–959. https://doi.org/10.1093/jn/128.6.954

Koh, K.K., Han, S.H., and Quon, M.J., 2005. Inflammatory markers and the metabolic syndrome: Insights from therapeutic interventions. *J. Am. Coll. Cardiol.*, 46(11): 1978–1985. https://doi.org/10.1016/j.jacc.2005.06.082

Kohlmeier, L., Kark, J.D., Gomez-Gracia, E., Martin, B.C., Steck, S.E., Kardinaal, A.F., Ringstad, J., Thamm, M., Masaev, V., Riemersma, R., Martin-Moreno, J.M., Huttunen, J.K., and Kok, F.J., 1997. Lycopene and myocardial infarction risk in the EURAMIC Study. *Am. J. Epidemiol.*, 146(8): 618–626. https://doi.org/10.1093/oxfordjournals.aje.a009327

Kucuk, O., Sarkar, F.H., Djuric, Z., Sakr, W., Pollak, M.N., Khachik, F., Banerjee, M., Bertram, J.S., and Wood, D.P., 2002. Effects of lycopene supplementation in patients with localized prostate cancer. *Exp. Biol. Med. (Maywood, N.J.)*, 227(10): 881–885. https://doi.org/10.1177/153537020222701007

Kucuk, O., Sarkar, F.H., Sakr, W., Djuric, Z., Pollak, M.N., Khachik, F., Li, Y.W., Banerjee, M., Grignon, D., Bertram, J.S., Crissman, J.D., Pontes, E.J., and Wood, D.P., 2001. Phase II randomized clinical trial of lycopene supplementation before radical prostatectomy. *Cancer Epidemiol. Biomarkers Prev.* 10(8): 861–868.

Lacombe, A., Li, R.W., Klimis-Zacas, D., Kristo, A.S., Tadepalli, S., Krauss, E., Young, R., and Wu, V.C.H., 2013. Lowbush wild blueberries have the potential to modify gut microbiota and xenobiotic metabolism in the rat colon. *PloS One*, 8(6): e67497. https://doi.org/10.1371/journal.pone.0067497

Lee, K.Y., Kim, J.-R., and Choi, H.C., 2016. Genistein-induced LKB1-AMPK activation inhibits senescence of VSMC through autophagy induction. *Vasc. Pharmacol.*, 81, 75–82. https://doi.org/10.1016/j.vph.2016.02.007

Lee, S., Keirsey, K.I., Kirkland, R., Grunewald, Z.I., Fischer, J.G., and de La Serre, C.B., 2018. Blueberry supplementation influences the gut microbiota, inflammation, and insulin resistance in high-fat-diet-fed rats. *J. Nutr.*, 148(2): 209–219. https://doi.org/10.1093/jn/nxx027

Li, S.-H., Liu, X.-X., Bai, Y.-Y., Wang, X.-J., Sun, K., Chen, J.-Z., and Hui, R.-T., 2010. Effect of oral isoflavone supplementation on vascular endothelial function in postmenopausal women: A meta-analysis of randomized placebo-controlled trials. *Am. J. Clin. Nutr.*, 91(2): 480–486. https://doi.org/10.3945/ajcn.2009.28203

Li, X., Jia, P., Huang, Z., Liu, S., Miao, J., Guo, Y., Wu, N., and Jia, D., 2019. Lycopene protects against myocardial ischemia-reperfusion injury by inhibiting mitochondrial permeability transition pore opening. *Drug Des. Devel. Ther.*, 13, 2331–2342. https://doi.org/10.2147/DDDT.S194753

Li, X. and Xu, J., 2013. Lycopene supplement and blood pressure: An updated meta-analysis of intervention trials. *Nutrients*, 5(9): 3696–3712. https://doi.org/10.3390/nu5093696

Li, Y., Zhang, Y., Liu, X., Wang, M., Wang, P., Yang, J., and Zhang, S., 2018. Lutein inhibits proliferation, invasion and migration of hypoxic breast cancer cells via downregulation of HES1. *Int. J. Oncol.*, 52(6): 2119–2129. https://doi.org/10.3892/ijo.2018.4332

Liu, D., Homan, L.L., and Dillon, J.S., 2004. Genistein acutely stimulates nitric oxide synthesis in vascular endothelial cells by a cyclic adenosine 5′-monophosphate-dependent mechanism. *Endocrinology*, 145(12): 5532–5539. https://doi.org/10.1210/en.2004-0102

Luo, T., Miranda-Garcia, O., Sasaki, G., Wang, J., and Shay, N.F., 2018. Genistein and daidzein decrease food intake and body weight gain in mice, and alter LXR signaling *in vivo* and *in vitro*. *Food Funct.*, 9(12): 6257–6267. https://doi.org/10.1039/c8fo01718b

Luo, T., Snyder, S.M., Zhao, B., Sullivan, D.K., Hamilton-Reeves, J., Guthrie, G., Ricketts, M.-L., Shiverick, K.T., and Shay, N., 2016. Gene expression patterns are altered in athymic mice and metabolic syndrome factors are reduced in C57BL/6J mice fed high-fat diets supplemented with soy isoflavones. *J. Agric. Food Chem.*, 64(40): 7492–7501. https://doi.org/10.1021/acs.jafc.6b03401

Messina, M.J. (2002). Soy foods and soybean isoflavones and menopausal health. *Nutrition in Clinical Care: An Official Publication of Tufts University*, 5(6): 272–282. https://doi.org/10.1046/j.1523-5408.2002.05602.x

Misra, R., Mangi, S., Joshi, S., Mittal, S., Gupta, S.K., and Pandey, R.M., 2006. LycoRed as an alternative to hormone replacement therapy in lowering serum lipids and oxidative stress markers: A randomized controlled clinical trial. *J. Obstet. Gynaecol. Res.*, 32(3): 299–304. https://doi.org/10.1111/j.1447-0756.2006.00410.x

Miyamoto, K., Hasuike, S., Kugo, H., Sukketsiri, W., Moriyama, T., and Zaima, N., 2022. Administration of isoflavone attenuates ovariectomy-induced degeneration of aortic wall. *J. Oleo Sci.*, 71(6): 889–896. https://doi.org/10.5650/jos.ess22043

Moradi, M., Daneshzad, E., and Azadbakht, L., 2020. The effects of isolated soy protein, isolated soy isoflavones and soy protein containing isoflavones on serum lipids in postmenopausal women: A systematic review and meta-analysis. *Crit. Rev. Food Sci. Nutr.*, 60(20): 3414–3428. https://doi.org/10.1080/10408398.2019.1689097

Mulero, J., Abellán, J., Zafrilla, P., Amores, D., and Hernández Sánchez, P., 2015. Bioactive substances with preventive effect in cardiovascular diseases. *Nutricion Hospitalaria*, 32(4): 1462–1467. https://doi.org/10.3305/nh.2015.32.4.9510

Nahum, A., Zeller, L., Danilenko, M., Prall, O.W.J., Watts, C.K.W., Sutherland, R.L., Levy, J., and Sharoni, Y., 2006. Lycopene inhibition of IGF-induced cancer cell growth depends on the level of cyclin D1. *Eur. J. Nutr.*, 45(5): 275–282. https://doi.org/10.1007/s00394-006-0595-x

Nishimura, M., Tominaga, N., Ishikawa-Takano, Y., Maeda-Yamamoto, M., and Nishihira, J., 2019. Effect of 12-week daily intake of the high-lycopene tomato (Solanum lycopersicum), a variety named "PR-7", on lipid metabolism: A randomized, double-blind, placebo-controlled, parallel-group study. *Nutrients*, 11(5): 1177. https://doi.org/10.3390/nu11051177

Norton, C., Kalea, A.Z., Harris, P.D., and Klimis-Zacas, D.J., 2005. Wild blueberry-rich diets affect the contractile machinery of the vascular smooth muscle in the Sprague-Dawley rat. *J. Med. Food*, 8(1): 8–13. https://doi.org/10.1089/jmf.2005.8.8

Obermüller-Jevic, U.C., Olano-Martin, E., Corbacho, A.M., Eiserich, J.P., van der Vliet, A., Valacchi, G., Cross, C.E., and Packer, L., 2003. Lycopene inhibits the growth of normal human prostate epithelial cells *in vitro*. *J. Nutr.*, 133(11): 3356–3360. https://doi.org/10.1093/jn/133.11.3356

Ohkura, Y., Obayashi, S., Yamada, K., Yamada, M., and Kubota, T., 2015. S-equol partially restored endothelial nitric oxide production in isoflavone-deficient ovariectomized rats. *J. Cardiovasc. Pharmacol.*, 65(5): 500–507. https://doi.org/10.1097/FJC.0000000000000220

Paran, E., Novack, V., Engelhard, Y.N., and Hazan-Halevy, I., 2009. The effects of natural antioxidants from tomato extract in treated but uncontrolled hypertensive patients. *Cardiovasc. Drugs Ther.*, 23(2): 145–151. https://doi.org/10.1007/s10557-008-6155-2

Peng, C., Gao, C., Lu, D., Rosner, B.A., Zeleznik, O., Hankinson, S.E., Kraft, P., Eliassen, A.H., and Tamimi, R.M., 2021. Circulating carotenoids and breast cancer among high-risk individuals. *Am. J. Clin. Nutr.*, 113(3): 525–533. https://doi.org/10.1093/ajcn/nqaa316

Peng, S.J., Li, J., Zhou, Y., Tuo, M., Qin, X.X., Yu, Q., Cheng, H., and Li, Y. M., 2017. *In vitro* effects and mechanisms of lycopene in MCF-7 human breast cancer cells. *Genet. Mol. Res.*, 16(2). https://doi.org/10.4238/gmr16029434

Peraita-Costa, I., Carrillo Garcia, P., and Morales-Suárez-Varela, M., 2022. Is there an association between β-carotene and breast cancer? A systematic review on breast cancer risk. *Nutr. Cancer*, 74(1): 39–54. https://doi.org/10.1080/01635581.2020.1865422

Petyaev, I.M., 2016. Lycopene deficiency in ageing and cardiovascular disease. *Oxid. Med. Cell. Longev.*, 1–6. https://doi.org/10.1155/2016/3218605

Petyaev, I.M., Dovgalevsky, P.Y., Klochkov, V.A., Chalyk, N.E., Pristensky, D.V., Chernyshova, M.P., Udumyan, R., Kocharyan, T., Kyle, N.H., Lozbiakova, M.V., and Bashmakov, Y.K., 2018. Effect of lycopene supplementation on cardiovascular parameters and markers of inflammation and oxidation in patients with coronary vascular disease. *Food Sci. Nutr.*, 6(6): 1770–1777. https://doi.org/10.1002/fsn3.734

Räthel, T.R., Leikert, J.F., Vollmar, A.M., and Dirsch, V.M., 2005. The soy isoflavone genistein induces a late but sustained activation of the endothelial nitric oxide-synthase system *in vitro*. *Br. J. Pharmacol.*, 144(3): 394–399. https://doi.org/10.1038/sj.bjp.0706075

Rodriguez-Mateos, A., Istas, G., Boschek, L., Feliciano, R.P., Mills, C.E., Boby, C., Gomez-Alonso, S., Milenkovic, D., and Heiss, C., 2019. Circulating anthocyanin metabolites mediate vascular benefits of blueberries: Insights from randomized controlled trials, metabolomics, and nutrigenomics. *J. Gerontol. A Biol. Sci. Med. Sci.* 74(7): 967–976. https://doi.org/10.1093/gerona/glz047

Rodriguez-Mateos, A., Rendeiro, C., Bergillos-Meca, T., Tabatabaee, S., George, T.W., Heiss, C., and Spencer, J.P., 2013. Intake and time dependence of blueberry flavonoid-induced improvements in vascular function: A randomized, controlled, double-blind, crossover intervention study with mechanistic insights into biological activity. *Am. J. Clin. Nutr.*, 98(5): 1179–1191. https://doi.org/10.3945/ajcn.113.066639

Ruiz Rejón, F., Martín-Peña, G., Granado, F., Ruiz-Galiana, J., Blanco, I., and Olmedilla, B., 2002. Plasma status of retinol, alpha- and gamma-tocopherols, and main carotenoids to first myocardial infarction: Case control and follow-up study. *Nutrition (Burbank, Los Angeles County, Calif.)*, 18(1): 26–31. https://doi.org/10.1016/s0899-9007(01)00683-9

Ruscica, M., Pavanello, C., Gandini, S., Gomaraschi, M., Vitali, C., Macchi, C., Morlotti, B., Aiello, G., Bosisio, R., Calabresi, L., Arnoldi, A., Sirtori, C.R., and Magni, P., 2018. Effect of soy on metabolic syndrome and cardiovascular risk factors: A randomized controlled trial. *Eur. J. Nutr.*, 57(2): 499–511. https://doi.org/10.1007/s00394-016-1333-7

Sakamoto, Y., Kanatsu, J., Toh, M., Naka, A., Kondo, K., and Iida, K., 2016. The dietary isoflavone daidzein reduces expression of pro-inflammatory genes through PPARα/γ and JNK pathways in adipocyte and macrophage co-cultures. *PloS One*, 11(2): e0149676. https://doi.org/10.1371/journal.pone.0149676

Sathyapalan, T., Aye, M., Rigby, A.S., Thatcher, N.J., Dargham, S.R., Kilpatrick, E.S., and Atkin, S.L., 2018. Soy isoflavones improve cardiovascular disease risk markers in women during the early menopause. *Nutr. Metab. Cardiovasc. Dis.*, 28(7): 691–697. https://doi.org/10.1016/j.numecd.2018.03.007

Sharifi-Zahabi, E., Soltani, S., Malekahmadi, M., Rezavand, L., Clark, C.C.T., and Shidfar, F., 2022. The effect of lycopene supplement from different sources on prostate specific antigen (PSA): A systematic review and meta-analysis of randomized controlled trials. *Complement. Ther. Med.*, 64, 102801. https://doi.org/10.1016/j.ctim.2022.102801

Shoker, R.M.H., 2020. A review article: The importance of the major groups of plants secondary metabolism phenols, alkaloids, and terpenes. *Int. J. Res. Appl. Sci. Biotechnol.*, 7(5): 354–358. https://doi.org/10.31033/ijrasb.7.5.47

Si, H., and Liu, D., 2008. Genistein, a soy phytoestrogen, upregulates the expression of human endothelial nitric oxide synthase and lowers blood pressure in spontaneously hypertensive rats. *J. Nutr.*, 138(2): 297–304. https://doi.org/10.1093/jn/138.2.297

Siler, U., Barella, L., Spitzer, V., Schnorr, J., Lein, M., Goralczyk, R., and Wertz, K., 2004. Lycopene and vitamin E interfere with autocrine/paracrine loops in the Dunning prostate cancer model. *FASEB J.*, 18(9): 1019–1021. https://doi.org/10.1096/fj.03-1116fje

Soares, N., da, C.P., Machado, C.L., Trindade, B.B., Lima, I.C. do, C., Gimba, E.R.P., Teodoro, A.J., Takiya, C., and Borojevic, R., 2017. Lycopene extracts from different tomato-based food products induce apoptosis in cultured human primary prostate cancer cells and regulate TP53, Bax and Bcl-2 transcript expression. *Asian Pac. J. Cancer Prev.*, 18(2): 339–345. https://doi.org/10.22034/APJCP.2017.18.2.339

Sowmya Shree, G., Yogendra Prasad, K., Arpitha, H.S., Deepika, U.R., Nawneet Kumar, K., Mondal, P., and Ganesan, P., 2017. β-carotene at physiologically attainable concentration induces apoptosis and down-regulates cell survival and antioxidant markers in human breast cancer (MCF-7) cells. *Mol. Cell. Biochem.*, 436(1–2): 1–12. https://doi.org/10.1007/s11010-017-3071-4

Squadrito, F., Altavilla, D., Squadrito, G., Saitta, A., Cucinotta, D., Minutoli, L., Deodato, B., Ferlito, M., Campo, G.M., Bova, A., and Caputi, A.P., 2000. Genistein supplementation and estrogen replacement therapy improve endothelial dysfunction induced by ovariectomy in rats. *Cardiovasc. Res.*, 45(2): 454–462. https://doi.org/10.1016/s0008-6363(99)00359-4

Stote, K.S., Wilson, M.M., Hallenbeck, D., Thomas, K., Rourke, J.M., Sweeney, M.I., Gottschall-Pass, K.T., and Gosmanov, A.R., 2020. Effect of blueberry consumption on cardiometabolic health parameters in men with type 2 diabetes: An 8-week, double-blind, randomized, placebo-controlled trial. *Curr. Dev. Nutr.*, 4(4): nzaa030. https://doi.org/10.1093/cdn/nzaa030

Suganami, T., Nishida, J., and Ogawa, Y., 2005. A paracrine loop between adipocytes and macrophages aggravates inflammatory changes: Role of free fatty acids and tumor necrosis factor alpha. *Arterioscler. Throm. Vasc. Biol.*, 25(10): 2062–2068. https://doi.org/10.1161/01.ATV.0000183883.72263.13

Suganami, T., Tanimoto-Koyama, K., Nishida, J., Itoh, M., Yuan, X., Mizuarai, S., Kotani, H., Yamaoka, S., Miyake, K., Aoe, S., Kamei, Y., and Ogawa, Y., 2007. Role of the toll-like receptor 4/NF-kappaB pathway in saturated fatty acid-induced inflammatory changes in the interaction between adipocytes and macrophages. *Arterioscler. Throm. Vasc. Biol.*, 27(1): 84–91. https://doi.org/10.1161/01.ATV.0000251608.09329.9a

Sung, H., Ferlay, J., Siegel, R.L., Laversanne, M., Soerjomataram, I., Jemal, A., and Bray, F., 2021. Global Cancer Statistics 2020: GLOBOCAN estimates of incidence and mortality worldwide for 36 cancers in 185 countries. *CA Cancer J. Clin.*, 71(3): 209–249. https://doi.org/10.3322/caac.21660

Tang, L., Jin, T., Zeng, X., and Wang, J.-S., 2005. Lycopene inhibits the growth of human androgen-independent prostate cancer cells *in vitro* and in

BALB/c nude mice. *J. Nutr.*, 135(2): 287–290. https://doi.org/10.1093/jn/135.2.287

Tong, C., Peng, C., Wang, L., Zhang, L., Yang, X., Xu, P., Li, J., Delplancke, T., Zhang, H., and Qi, H., 2016. Intravenous administration of lycopene, a tomato extract, protects against myocardial ischemia-reperfusion injury. *Nutrients*, 8(3): 138. https://doi.org/10.3390/nu8030138

Torres, N., Torre-Villalvazo, I., and Tovar, A.R., 2006. Regulation of lipid metabolism by soy protein and its implication in diseases mediated by lipid disorders. *J. Nutr. Biochem.*, 17(6): 365–373. https://doi.org/10.1016/j.jnutbio.2005.11.005

Upaganlawar, A., Gandhi, H., and Balaraman, R., 2010. Effect of vitamin E alone and in combination with lycopene on biochemical and histopathological alterations in isoproterenol-induced myocardial infarction in rats. *J. Pharmacol. Pharmacother.*, 1(1): 24–31. https://doi.org/10.4103/0976-500X.64532

Van Hoang, D., Pham, N.M., Lee, A.H., Tran, D.N., and Binns, C.W., 2018. Dietary carotenoid intakes and prostate cancer risk: A case-control study from Vietnam. *Nutrients*, 10(1): E70. https://doi.org/10.3390/nu10010070

Vedavanam, K., Srijayanta, S., O'Reilly, J., Raman, A., and Wiseman, H., 1999. Antioxidant action and potential antidiabetic properties of an isoflavonoid-containing soyabean phytochemical extract (SPE). *Phytother. Res.*, 13(7): 601–608. https://doi.org/10.1002/(SICI)1099-1573(199911)13:7<601::AID-PTR550>3.0.CO;2-O

Velasquez, M.T., and Bhathena, S.J., 2007. Role of dietary soy protein in obesity. *Int. J. Med. Sci.* 4(2): 72–82. https://doi.org/10.7150/ijms.4.72

Vendrame, S., Guglielmetti, S., Riso, P., Arioli, S., Klimis-Zacas, D., and Porrini, M., 2011. Six-week consumption of a wild blueberry powder drink increases bifidobacteria in the human gut. *J. Agric. Food Chem.*, 59(24): 12815–12820. https://doi.org/10.1021/jf2028686

Wang, Y., Gapstur, S.M., Gaudet, M.M., Furtado, J.D., Campos, H., and McCullough, M.L., 2015. Plasma carotenoids and breast cancer risk in the Cancer Prevention Study II Nutrition Cohort. *Cancer Causes Control*, 26(9): 1233–1244. https://doi.org/10.1007/s10552-015-0614-4

Weisberg, S.P., McCann, D., Desai, M., Rosenbaum, M., Leibel, R.L., and Ferrante, A.W., 2003. Obesity is associated with macrophage accumulation in adipose tissue. *J. Clin. Invest.*, 112(12): 1796–1808. https://doi.org/10.1172/JCI19246

Wilson, P.W.F., D'Agostino, R.B., Parise, H., Sullivan, L., and Meigs, J.B., 2005. Metabolic syndrome as a precursor of cardiovascular disease and type 2 diabetes mellitus. *Circulation*, 112(20): 3066–3072. https://doi.org/10.1161/CIRCULATIONAHA.105.539528

Wolak, T., Sharoni, Y., Levy, J., Linnewiel-Hermoni, K., Stepensky, D., and Paran, E., 2019. Effect of tomato nutrient complex on blood pressure: A double blind, randomized dose-response study. *Nutrients*, 11(5): E950. https://doi.org/10.3390/nu11050950

Woo, H.W., Kim, M.K., Lee, Y.-H., Shin, D.H., Shin, M.-H., and Choi, B.Y., 2019. Habitual consumption of soy protein and isoflavones and risk of metabolic syndrome in adults ≥ 40 years old: A prospective analysis of the Korean Multi-Rural Communities Cohort Study (MRCohort). *Eur. J. Nutr.*, 58(7): 2835–2850. https://doi.org/10.1007/s00394-018-1833-8

World Health Organization, 2011. Global status report on noncommunicable diseases 2010. World Health Organization. https://apps.who.int/iris/handle/10665/44579

Wu, K., Erdman, J.W., Schwartz, S.J., Platz, E.A., Leitzmann, M., Clinton, S.K., DeGroff, V., Willett, W.C., and Giovannucci, E., 2004. Plasma and dietary carotenoids, and the risk of prostate cancer: A nested case-control study. *Cancer Epidemiol. Biomarkers Prev.* 13(2): 260–269. https://doi.org/10.1158/1055-9965.epi-03-0012

Zhang, H., Zhao, J., Yu, H., and Guo, D. 2016. Genistein ameliorated endothelial nitric oxidase synthase uncoupling by stimulating sirtuin-1 pathway in ox-LDL-injured HUVECs. *Environ. Toxicol. Pharmacol.*, 42, 118–124. https://doi.org/10.1016/j.etap.2016.01.011

Zhang, H., Zhao, Z., Pang, X., Yang, J., Yu, H., Zhang, Y., Zhou, H., and Zhao, J., 2017. Genistein protects against ox-LDL-induced inflammation through microRNA-155/SOCS1-mediated repression of NF-κB signaling pathway in HUVECs. *Inflammation*, 40(4): 1450–1459. https://doi.org/10.1007/s10753-017-0588-3

Zhang, X., Spiegelman, D., Baglietto, L., Bernstein, L., Boggs, D.A., van den Brandt, P.A., Buring, J.E., Gapstur, S.M., Giles, G.G., Giovannucci, E., Goodman, G., Hankinson, S.E., Helzlsouer, K.J., Horn-Ross, P.L., Inoue, M., Jung, S., Khudyakov, P., Larsson, S.C., Lof, M. *et al.*, 2012. Carotenoid intakes and risk of breast cancer defined by estrogen receptor and progesterone receptor status: A pooled analysis of 18 prospective cohort studies. *Am. J. Clin. Nutr.*, 95(3): 713–725. https://doi.org/10.3945/ajcn.111.014415

https://doi.org/10.1142/9789811276736_0004

Chapter 4

The Importance of Seed Quality in International Seed Trade

Anthony B. Tse

Clover Seed Company Limited
atse@cloverseed.com.hk

1. Global Seed Trade

Seeds are living organism and essential in today's agriculture. Many of our food, oil, fibre and ornamental crops are propagated by seeds. The successful cultivation of these crops relies on seeds and particularly we need quality seeds. We need seeds of good genetic quality, high germination, disease free and high physical quality to cultivate a successful crop.

At the same time the seed trade is highly globalised and no longer remains an industry bound by national boundaries. Breeding of new and improved varieties is now depended on not only conventional breeding methods but also supplemented by modern biological technologies. This not only requires high and advance technology but at the same time, time consuming, and requires investment in equipment and human resources. It takes easily over ten years to develop a new variety so the return on investment is very long. These are concentrated in North America, Northern Europe and Japan by some very specialised companies or government institutes.

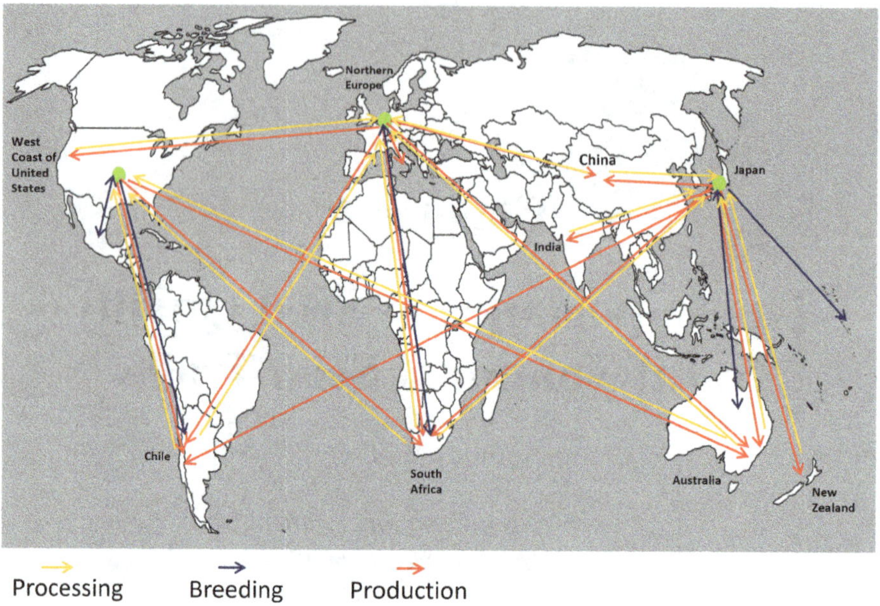

Figure 1. Global seed movement.

Seed production at the same time is not necessary where the variety is bred but at production areas where the climate is most favourable, disease free and with availability of skilled labour.

Seed processing is usually partially done at production site and then finish at the site of breeding. High quality seeds today demand sophisticated machineries and techniques to raise them to the high standard the trade demands today. The seeds are then stored in climate-controlled facilities and then shipped all around the world.

2. Seed Production Process

Seed production is a very specialised process and other than the traditional crops where farmers saved their own seeds, many of the commercially traded crops are produced in overseas locations where the climate is most favourable. Not only the conditions are needed to be most favourable, there must be a dry period to coincide with the harvesting and at the same time ample irrigation water. Many of these are of typical Mediterranean climatic areas like West Coast of United States, Chile, South Africa, some parts of New Zealand and Australia and countries around the Mediterranean Sea. These are all very important areas for opened field crops which can be mechanically planted and harvested.

Second to most favourable climatic conditions the production area must be free from seed transmitted diseases. We now know certain diseases are seed borne and can be transmitted through seeds and these are definitely not acceptable. These are considered quarantine pests and are not allowed to cross international borders.

For some hybrid crops the emasculation and pollination process requires hand labour and this is extremely labour intensive. These include Solanaceous crops like tomato, pepper and eggplant, Cucurbit crops like cucumber, squash, gourds and watermelon, and various flower crops. The hybrid seed production of these crops requires stable supply of skilled labours with sharp eyes and delicate fingers. Places where these skilled labours are available are limited and today, they are mainly in China, India, Chile, Thailand and Vietnam. If managed properly this is a very good way to improve livelihood of the local people, particularly creating employment opportunities for women.

As we can see the seed trade today is no longer confined to within national borders and it is not uncommon for one lot of seeds to cross several national borders before it reaches its final destination.

According to the International Seed Federation the total world seed export in 2020 is 7,265 tons or USD 15.85 million (ISF, 2022) and seed import the same year is 7,502 tons or USD14.92 million (ISF, 2022). Compare to the total world trade these are very small numbers and may be insignificant. However, seeds are only a small percentage of agricultural production, often less than 1% of total input. In this case, the effects of the international seed trade will be very significant.

3. Seed Qualities

Seed is one commodity that the quality and particularly its potential in agricultural yield cannot easily determined by its present state. It has often been subjected to fraud. Because of this and its importance the seed trade is highly regulated by governments and self-regulated by seed companies themselves to ensure the seeds meet the necessary qualities.

There are several aspects of seed quality that can be measured and these are germination, vigour, moisture content, and physical purity. Seed health can be often tested with standardised protocols, where as the genetic purity of the seeds are sometimes relied on government agencies to certify or solely the responsibility of the seed companies.

Seed germination — seeds are basically useless unless they can germinate. Not only they need to germinate and give a healthy plant, the percentage they will germinate is also important. Usually, the trade will accept minimum germination at 85% and more and more now the standard requirement is 90% or even 95%.

Seed vigour — this is now recognised as just as important as germination. Modern farming demands high uniform germination to give uniform seedlings or uniform stands. Abnormal seedlings, seed with poor radicals are counted out as they will not make normal useful seedlings. Uniform germination within a time frame is also considered necessary.

Moisture content — This is highly related to the shelf life of the seeds. Seeds are living organism and do have limited shelf life and this is highly related to their moisture content. Unless the seeds are dried to an acceptable moisture content, the shelf life will be greatly affected. This is usually need to be lower than 7% — 8%. Too low a moisture content on the other hand can damage the seeds.

Physical purity — seeds should be seeds only and nothing else. Soil particles, rocks, broken seeds and other inert materials should not be there. Of utmost importance is the presence of seeds of noxious weeds. These are quarantine pests and their presence may cause the seed lots to be destroyed. Normally the trade expects physical purity to be 99%.

Genetic purity — this is to assure the seeds supplied are what they are and not something else. This is probably the most difficult aspects of seed quality to be determined. As it is there is no easy way to determine this unless the seeds are grown out in a field test to determine and assure what they are. The use of molecular markers or DNA finger printing can be useful but this is not readily available yet. For many open pollinated crops there are certifying agencies to verify this. For Hybrids usually it is up to the seed company to assure this by doing grow out test or by the means of molecular markers. Depending on the crop the genetic purity is expected to be at least 98%.

Seed health — seeds are living organisms and can carry and transmit certain seed borne diseases. It is essential that this does not happen otherwise serious economic losses can occur and seed companies are legally liable for this. Not all diseases are transmitted through seeds, some of the more serious ones are:

- Black rot in brassicas
- Bacterial fruit blotch in Cucurbits
- Tomato brown rugose fruit virus in Solanaceae crops
- Bacterial canker in tomato
- Cucumber green mottle mosaic virus in cucurbits
- Candidatus liberibacter solanacearum in Solanaceae and Umbelliferae crops

For a complete list see ISF Pest List (ISF, n.d.)

Seed companies are responsible to make sure that their seeds do not carry such diseases by producing from clean stock seeds. Produce in known disease-free areas and take all necessary precautionary measures to prevent cross contamination. Also, some diseases can be effectively disinfected by physical or chemical means after harvest. Besides laboratory tests are available to determine the seeds are effectively disease free.

All these quality issues are the responsibility of the seed companies and it is their responsibility to make sue their products are up to standard. All honest and responsible seed companies will do so as they are liable not only to the cost of the seeds but the economic loss of the farmers. At the same time there are International Agencies that will certify these.

4. Certifications and International Agencies

International Seed Testing Association (ISTA) — founded in 1924, with the aim to develop and publish standard procedures in the field of seed testing. It is inextricably linked with the history of seed testing and has member laboratories in over 80 countries/distinct economies worldwide.

Its membership is truly a global network. Its vision is to establish a uniform system in seed quality evaluation world-wide. It produces internationally agreed rules for seed sampling and testing, accredits laboratories, promotes research, provides international seed analysis certificates and training, and disseminates knowledge in seed science and technology. This facilitates seed trading nationally and internationally and also contributes to food security (ISTA, n.d.).

ISTA has over 130 approved laboratories all over the world. These are accredited to issue Orange or Blue Certificates to certify the germination, moisture content and purity of the seeds. In additional they will issue certificate stating the seeds are free from specific seed borne disease. Many countries will require the importation of seeds to be accompanied by Orange or Blue Certificates.

The Organization for Economic Cooperation and Development OECD — this organization promotes the use of certified agriculture seeds. The OECD Schemes for the Varietal Certification of Seed or controlling of seeds traded internationally was established in 1958. Membership is opened to OCED, UN and WTO countries. The principle aims of all eight seed schemes which covers more that 200 agricultural and vegetable species, is to encourage the production and use of high-quality seeds. These seeds are produced — and officially controlled — according to a set of harmonised procedures put in place in the 61 participating countries. By ensuring consistently high standard, the schemes contribute to its members' evolving agriculture and trade policies (OECD, n.d.).

The United States is not part of OECD Seed Scheme. They have their own system of seed certification through the State Agriculture Extension Services. Its established system is to inspect seed production fields and certify the seeds are true to type and will issue certificates and tags for each bag of seeds. For crops outside of these schemes particularly Hybrid Vegetable crops it is up to individual companies to do their own laboratory or field grow out tests.

International Plant Protection Convention (IPPC) — is an intergovernmental treaty signed over 180 countries. Its aim is to protect the world's plant resources from the spread and introduction of pests and promoting safe trade. The Convention introduced International Standards for Phytosanitary Measures (ISPMs) as its main tool to achieve its goal, making it the sole global standard setting organization for plant health (IPPC, n.d.). The IPPC is one of the "Three Sisters" recognised by the World Trade Organization's (WTO) Sanitary and Phytosanitary Measures (SPS) Agreement, along with the CODEX Alimentarius Commission for food safety standards and the World Organization for Animal Health (OIE) for animal health standards. The National Plant Protection Officer (NPPO) of each member country will set up import requirements of seeds into their own country. They will issue Import Permits and Export Phytosanitary Certificates for the shipment of seeds. Such permits are essential for the movement of seeds across international borders.

In conclusion seeds are important and essential in today's agriculture production. Seeds are produced in areas best suited for their production and shipped all over the world. Seeds are needed to be of high quality to ensure successful production. The seed trade is governed and regulated by different international organizations. At the same time, it is also high self-regulated by the seed companies themselves, as their liability is not limited to the cost of the seeds but the economic loss of the farmers which can be many times more.

References

International Plant Protection Convention. (n.d.). Retrieved September 28, 2022, from www.ippc.int/zh/about/overview/

International Seed Federation. (n.d.). *ISF regulated pest list database*. Retrieved September 28, 2022, from www.pestlist.worldseed.org/public/pestlist.jsp

International Seed Federation. (2022). *SEED EXPORTS 2020*. Retrieved from https://worldseed.org/document/seed-exports-2020/

International Seed Federation. (2022). *SEED IMPORTS 2020*. Retrieved from https:// worldseed.org/document/seed-imports-2020/

International Seed Testing Association. (n.d.). *About Us*. Retrieved September 28, 2022, from www.seedtest.org/en/informations-footer/about-us.html

The Organization for Economic Cooperation and Development. (n.d.). Retrieved September 28, 2022, from www.oecd.org/

https://doi.org/10.1142/9789811276736_0005

Chapter 5

Soybean and Changing Food Culture in Chinese History

Angela Ki Che Leung

Hong Kong Institute for the Humanities and Social Sciences,
University of Hong Kong
kcleung7@hku.hk

Soybean, known as *shu* 菽 in Chinese classics, a native crop in China with multiple uses, was considered one of the five fundamental food crops in early Chinese agriculture, the other four being *ji* (稷setaria millet), *shu* (黍panicum millet), *dao* (稻rice), *mai* (麥wheat and barley). According to H.T. Huang who did a thorough study on traditional Chinese food technology, unlike the other four grains which could easily be cooked to be eaten as a staple food, soybean was no longer consumed as a staple food after the 6th century when more complicated processing techniques including prolonged fermentation, grinding, filtering, sprouting, etc. had transformed soybean into an ingredient for making tasty or nutritious non-staple foods, including condiments such as *jiang* paste, soy sauce, *shi* fermented beans, or substantial foods such as bean curd, yuba, or tofu skin. A good part of the technology for making these foods is recorded in *Qimin yaoshu* (齊民要術 Essential Techniques for the People) of the 6th century, one of the earliest and most important Chinese agricultural treatises. (Huang, 2000) Not being a staple food crop like rice or wheat after the Tang dynasty, soybean played a unique role in China's agronomy and food culture ever since. As the crop grew almost everywhere in China, though not in great quantity, it was also used as animal feed, and as a natural fertilising crop to rotate with staple crops, with its nitrogen-rich roots.

This chapter tells the story of how soybean as a non-staple but popular food crop shapes Chinese history since late imperial time. It highlights the legume's unique role in the creation of the empire-wide urban food market in China of the 18th century, and how it articulates Chinese vision of modernity in the early 20th century when China was facing increasing challenges from the West. The modern history of soybean in China prepares its global role today.

1. Soybean in Late Imperial Chinese Food Culture

Chinese traditional diet was mostly, though not exclusively, vegetarian. It was not necessarily a religious practice, nor one limited to the lower classes who could not afford animal food. Even the diet of the elite literati was essentially vegetarian by choice, a tradition that was already observable in the Song dynasty (960–1279). In the late imperial period, Li Yu (李漁 1611–1680), a leading literary figure, maverick scholar and Ming loyalist from the Jiangnan region, explained in his famous jottings, *Xianqing Ou ji* (閒情偶記 Leisure notes) on the beauty of vegetarian foods: "They are light, clean, sweet-smelling, crispy. But people do not understand that their greatest beauty lies in their fresh savour (*xian*鮮) that surpasses that of meats." He gave the example of wild bamboo shoots, a favourite vegetable for Jiangnan elites. "Boil them plainly and eat with a little soy sauce, as the most exquisite food is often prepared simply and consumed with nothing else." Soy sauce (*jiangyou* 醬油), for Li Yu was clearly the key that brought out the superior *xian* or "meaty" flavour in vegetables appreciated by food connoisseurs.

Soy sauce, one of the most popular Chinese condiments today was then enjoyed mostly by the elite classes. The first extant recipe for making soy sauce was recorded by the famous Yuan painter Ni Zan (倪瓚 1301–1374) who mentioned the key steps of fermentation of soybeans beginning in the hot season and brewing in brine. More than two centuries later, a Ming literatus Tian Yiheng (田藝蘅 1524–1591), son of a prominent Zhejiang scholar-official told his readers that various soy foods, including soy sauce were "rare foods for modest peasants. Those living in isolated hills and valleys might never taste it in their whole life", and that products that one could buy in the market were often corrupt. In other words, soy foods were still enjoyed only by the privileged classes, even in prosperous regions in Ming China (Leung, forthcoming)

The example of soy sauce was only one of the many examples of the slow popularisation of soy foods despite the fact that the technology of making them had been known for centuries. In some of the late Ming and early Qing compilations of recipes, one can see that elite families were making their own soy foods, such as pastes (*jiang* 醬) and sauce, fermented beans, bean curds, and yoba (skin of bean curds), and probably other soy foods which were not yet widely commodified. The wider society was able to enjoy the foods more easily only when these became affordable and increasingly commodified from the 18th century onward. The key to this change was the Manchu conquest of China in 1644 allowing the full integration of Manchuria in China luring continuous waves of migration of Chinese peasants, transforming the vast region into a major agricultural zone in the Qing empire specialised in the production of soybean as a cash crop. Before this moment, soybeans in China, though grown almost everywhere, were cultivated mostly on road-sides, on fringes of fields, or as a gap crop rotating with staple crops like rice, largely insufficient to provide for the market of the empire.

2. Manchuria and the Popularisation of Soy Foods in the Qing Empire

Manchuria before the Qing conquest of China in 1644 was culturally and economically distinct from China proper. It was not an agricultural region, but mostly pastoral land and forests where Jurchen and Mongol nomads thrived on hunting, cattle rearing, gathering of forest products, fishery, etc. Soybeans were indigenous crops, but grown in relatively small quantities used mostly as fodder for horses and camels and traditionally levied in kind as a land tax until the late 17th century. The situation changed when the region became part of the Qing empire, allowing Han Chinese peasants to move in and settle to cultivate land, which was not the traditional occupation of the Manchu. Soybean soon became the most lucrative crop to grow in the region, used not only as an animal feed as in the Manchu tradition, but mostly for exportation to China proper on a massive scale as ingredient for human soy foods, oil making, or in the form of soy meal as fodder or fertiliser. (Yi, 1993; Diao and Yi, 1994; Ding *et al.*, 2004)

The first Qing emperor Shunzhi (1644–1661) already saw the need to develop agriculture in Manchuria to secure food provision for its population and the military units of the Manchu home base. In 1653, he announced the first edict to encourage the immigration of Chinese peasants to cultivate vast Manchurian land, by providing them with loans in the form of grain and buffalo — twenty heads for every hundred peasants. This announcement triggered the first wave of Chinese peasants moving into Manchuria, followed by successive waves throughout the dynasty, all aspiring to made a good living by cultivating lucrative soybean. (Yi, 1993) Before the mid-18th century however, for fear of jeopardising food security in Manchuria, the Manchu state only allowed the beans to be exported to China proper in limited quantities via land routes and along the Grand Canal. (Kato, 1953) The increasing but still insufficient supply of Manchurian beans in China proper paradoxically exposed ever more clearly its long-standing soybean shortage. The problem of shortage was fully explained in a memorial issued in 1763 by Yang Tingzhang, governor-general of Fujian-Zhejiang provinces, in which he complained about the insufficient supply of Manchurian beans: "Yellow soybeans are used by the people in making *jiang* condiments, tofu, vegetarian dishes, and as fodder for horses. There is a huge demand. But Zhejiang soybean production is inadequate, and the amount received from Fengtian (Manchuria) is small … and does not reach all counties." Yang pleaded for direct maritime shipments of soybeans from Shandong province (where production was also abundant), in hopes that the trade would "bring down the price of the beans (in Zhejiang). This would be greatly beneficial to the livelihood of people in the province…" His memorial shows that the limited access to Manchurian soybean in the Chinese market in the mid-18th century only served to whet the Chinese appetite for more soybean (Leung, forthcoming).

It was only from the early 1770s, when the Manchu state felt politically and militarily secure, and realised the great fiscal benefits of taxing soybean exports from Manchuria that the ban on maritime trade along the Chinese coast was progressively lifted. This policy allowing bulk shipments of beans on big boats descending China's eastern coastline greatly boosted the supply in China proper. The growing economic value of soybean cultivation encouraged more Chinese peasants to move into Manchuria until the early 20th century when northern Manchuria was also open to agriculture and immigration, further expanded the production and exportation of soybeans, not only to China proper, but also to the world. China thus became a global soybean producer and exporter (Kato, 1953; Yi 1993; Ding *et al.*, 2004; Kung and Li 2011).

The increase of Manchurian soybean commodification and exportation to China has been well studied: While the annual Manchurian soybean surplus was under 1.5 million *shi* (石, around 118,800 tons) in the late 17th century, it increased to 3 million *shi* in the first half of the 18th century, topped 4 million *shi* in 1875 and reached 7 million in 1900. Estimated exports from Manchuria to China proper increased from 1 to 1.5 million *shi* in the late 18th century to 3 million in 1875, and more than 5 million by 1900 (Isett, 2007). One consequence of the thriving maritime soy trade was the spectacular rise of Shanghai, ultimately overshadowing Suzhou as the leading port city in Jiangnan in the 18th century. The Shanghai soy trade guild (*douhang* 豆行) established in 1765 was a landmark, and by the mid-19th century, the guild at Cuixiu Hall (萃秀堂), in the City God Temple Garden became the biggest and most powerful of the 21 merchant guilds in the city by 1843, and a key stakeholder of the city's governance. Its site is still a tourist attraction today. Manchurian soybean traders along China's eastern coast formed a tight network and subsequently acquired a dominant position in municipal governance in the most prosperous part of China in the late imperial period (Fan, 2013).

The continuous and ample supply of Manchurian soybeans explains the popularization of culinary practices based on the everyday use of soy-based condiments that one can find in popular Qing recipe books. The emergence and subsequent strong presence of sizable urban soy food manufactures (or pickle shops — *jiangyuan* 醬園, or *jiangfang* 醬坊) in the urban landscape was a mid-Qing phenomenon. These shops made all kinds of fermented foods and drinks, mostly with soy sauce as the branding product. A 1933 gazetteer defined the function of pickle shops as follows: "They can make anything that regulates taste, especially soy sauce". In official documents and literati writings, one can find sporadic mentions of soy manufactures in North China, the Jiangnan region, and coastal cities, beginning in the Yongzheng period (1723–1735) and increasingly throughout the 18th and 19th centuries. Pickle shops in major maritime and riverine trade ports grew in significant numbers and began to establish collective guilds (*gongsuo* 公所) in the 1870s. The earliest, and one of the biggest, was the one established in 1873 in Suzhou by 86 pre-existing local shops. By the early 20th century, most counties had more than one shop, and some of the recorded manufacturers claimed their origin in the earlier Qing period or even in the late Ming. By the end of the Qing empire, pickle shops had become a major component of the Chinese urban landscape, some of which were identified with the cities themselves, such as Liubiju (六必居) for Peking, Yutang (玉堂) for Jining (Shandong), Hu Yumei (胡玉美) for Anqing (Anhui), Dingfeng (鼎豐) for Pinghu (Zhejiang), Feng Wantong (馮萬通) for Shanghai, with emerging brands in southern provinces like Hunan, Sichuan, Fujian, and Guangdong (Leung, forthcoming).

Republican sources provide more precise information that depicts the magnitude of the pickle shop phenomenon as an important marker of late imperial Chinese urban food culture. A 1926 article on Peking pickle shops gave a count of more than 140 shops of different sizes, while a 1933 report on the industry in Zhejiang gave a count of 322 pickle shops in the province with larger concentrations of over 80 shops in two counties and anything between a dozen to over 50 shops for the other counties. Many of these shops were established in the Qing period. Zhao Ronguang (趙榮光), food historian in China, did a first systematic study of Shanghai pickle shops in 2005, based on shops registered in the city in the 1930s and 1940s. He identified the names and addresses of 102 shops of the "traditional" artisanal type, and 251 "modern" shops using chemical methods. This counting, which Zhao still considered incomplete, does show that in Shanghai alone, more than one hundred old traditional pickle shops were doing business in the early 20th century, many of which were probably established in the Qing empire. Some of the allegedly "modern" shops might also have been established earlier and changed production methods in the Republican period (Zhao, 2005). The data fully reflects the burgeoning of an urban food culture in the Qing empire and early Republican China surrounding the proliferation of pickle shops from the late 18th century, made possible by the ample supply of Manchurian soybeans.

3. Soybean and Chinese Modernity

Soybean with its growing economic and cultural importance during the Qing empire acquired yet other values at the turn of the 20th century. It was a critical moment when China, facing unprecedented challenges from the West, was envisioning a modernity defined by Western ideals but also taking into account its own history. Chinese revolutionaries and political idealists soon identified soybean, a traditional and indigenous food crop, as the symbol of Chinese dietetic modernity, when western scientific nutritional knowledge was introduced.

Li Shizeng (李石曾alias Li Yuying李煜瀛, 1881–1973), an early anarchist from a prominent scholar-official family, was one of the first pioneer soybean scientists who studied and published the chemistry of soybean in France during the first decade of the 20th century when he also set up a factory in a Parisian suburb to make soy foods, including tofu and soy milk. Li had a much bigger agenda behind his early efforts in studying soybean as an intellectual pursuit. His plan was to demonstrate to the world the superiority of Chinese dietetics, and China's readiness to be part of the modern world. He joined the Republican revolutionary party led by Sun Yat-sen (1866–1925) in 1907 and published many articles on soybean and vegetarian diets thereafter. In 1908 he wrote an article to prophesise that Chinese traditional tofu as a cheap nutritious food would be widely accepted by the rest of the world in the new 20th century and recommended vegetarian diet for the modern world. As a westernised intellectual, Li highlighted the scientific and secular aspects of vegetarianism, distancing himself from Buddhist vegetarianism . He wrote that vegetables were better foods because they contained all the basic nutrients (protein, fat, and especially carbohydrates) and were less toxic than meat. Li's most important conclusion on Chinese vegetarianism was that since food plants were cheaper and better sources of nutrients, China should encourage its people to retain its vegetarian dietetic culture and not be misled to follow the Western model of developing meat industries. The choice of soybean as his research topic in France cannot be better understood in this context: as a legume known for its high protein content, it was for Li the ideal food crop for a poor, but modernising China.

Li's influence on Sun Yat-sen, who established the Republic of China after overthrowing the Qing regime in 1911, can be seen in Sun's plans for building the modern Chinese Republic as drawn out in his *Strategies in State Building* (1919). The text began with a chapter on "Food and Drink" in which he praised China for its dietetic traditions which, to him, were "superior to all civilised countries", contrary to China's other areas which were "backward" compared to the West. He mentioned in particular Chinese superior techniques in making vegetarian foods. Compared to Western vegetarian dishes which were, for him, not refined nor sufficiently nutritious, "in China, vegetarians eat bean curd, which is actually the 'meat' in vegetarian diets... This is why China is entirely attuned to vegetarian diets even before (Western) scholars made recommendation." He was clearly convinced by Li and other vegetarian political activists of his time that Republican China as a modern state should not follow the Western example of developing meat industries, given the country's huge soybean resources and its strong vegetarian tradition (Leung, 2019; Fu, 2018).

However, modern, secular vegetarianism did not become the "national diet" for Republican China despite the efforts of Sun and Li, and was severely challenged by US-trained scientists, like Wu Xian (吳憲 1893–1959) whose studies on vegetarian diet in the 1920s established the link between the diet, found to be dangerously deficient, and China's "inferior culture". This position was shared by many Chinese intellectuals of the time recommending thorough westernization for China to reinvigorate its economic and cultural strengths (Leung, 2019).Yet at the same time of the developing controversy, soybean, increasingly seen as an important food and industrial crop, became a subject of global scientific research. Chinese agronomists trained in Japan or the US began to carry out research in experiment stations or in major universities to develop new high-yield breeds to meet increasing market demand. The task of improving Chinese soybean breeds was, for Chinese agronomists, like Wang Shou (王綬 1897–1972), Cornell-trained agronomist, a patriotic act above all, as "soybean is China's lifeline", being China's most nutritious food plant, accounting for 21% of China's total export (in 1934). In 1933 he gave a lecture at Cornell on the future of Chinese soybean in the US. After providing the scientific data on soybean, especially its high protein content, he explained to his American audience that soybean was the most cost-effective way for the Chinese to obtain their required nutrition as they ingested the protein directly by eating soy foods, and not indirectly by eating the meat of the cattle fed on soybean, which was a much more costly procedure that China could hardly afford. He continued his prediction for the future of soybean in the US, "As the population in your country is still sparse, and you have vast spans of farmland, you are natural meat-eaters. But I truly think that when your population grows, you will change your diet from animal-based to vegetarian. This will certainly happen to your people, as it has happened in China. You will then realise the value of soybean".

While promoting the dietetic value of soybean in his alma mater echoing Li Shizeng's recommendations some thirty years before, Wang Shou was certainly aware of the imminent crisis that China was about to face when soybean cultivation in Manchuria was reaching its limits after more than two centuries of cultivation with traditional knowledge and technology. His Cornell lecture took place just before the upward trajectory of Manchurian bean production since the 18th century was about to reverse. A 1936 article in New York Times on soybean as the rising star of American agriculture clearly indicates the forthcoming, almost irreversible trend "Soybeans were the crop which earned the American farmer a $40,000,000 income last year on a five and three-quarter million acreage. "Soybean" was the name which a few weeks ago went up on the "big board" in the Chicago Board of Trade's latest listing...(soybean) in its new phases of the 1930s seems the entering wedge for a thriving market for American farm products in American industry". American advanced agronomic science experimented on vast and sparsely populated stretches of farm land quickly made the country into the world's leading producer of soybean, replacing China in 1954 (Wang and Guo, 2007).

The rise of American soybeans marked the beginning of a long odyssey for Chinese agronomists like Wang and his peers to regain soybean self-sufficiency for China by boosting productivity with new science. During the first decades of the 20th century, the technique of cross breeding soybeans for creating new high-yield cultivars was mastered mostly by Japanese experts in experiment stations in Manchuria and US scientists in America. Wang himself was struggling to develop new breeds in the Nanking region in the 1920s and 1930s applying what he had learned in Cornell to very different local situations and regional cultigens in China, also having to work with insufficient public institutional support under a weak government, and with peasants who were unfamiliar with the science. The efforts were then interrupted by wars, revolutions, and political instability that further delayed concerted efforts until the late 1970s. From the 1980s until today, soybean improvement in China remains to be driven primarily by the need of productivity increase. It is still an enormous uphill struggle: in 2007, the average productivity per *mu* in China was 96.9 kg, which was 55.1 kg below world average. In 2020, China imported more than 100 M tonnes, accounting for 85% of the country's domestic consumption that year and about 60% of the global soybean trade volume. Under such immense pressure, pushing up productivity will certainly continue to be the main target of national soybean research programs. (*South China Morning Post*, 14th January 2022)

There are, however, recent signs that the re-cultivation of soybeans has acquired other meanings than yield increase. Some of the efforts are now applied to the re-cultivation of indigenous strains for the making of heritage soy foods, often deployed in collaboration with local soy food makers and agrobusinesses, a development similar to black bean re-cultivation in Taiwan for the making of a unique type of soy sauce and health foods since the 1990s. Such efforts encourage the participation of wider public sectors and local businesses, as they also serve to preserve agronomic memory by reviving indigenous cultigens, and thus biodiversity, as well as the promotion of local, heritage foods largely defined by the localness of their ingredients, such as the crops. Such movements have already been growing steadily in Japan and South Korea before they started in Taiwan. When soybean is appreciated not only as a money-making cash crop, but especially as an ingredient that defines foods that expresses our identity and culture, the interest of the public in its re-cultivation could be immeasurable, and the impact beyond the economics. (Lian, 1994) The intensity and depth of popular food culture based in no small part on soybean since the 18th century is a strong historical memory. It would be interesting to see if the heritagization of crops and foods in this post-industrial era will breathe new life into this ancient Chinese food crop with unique traits (Du Bois, Tan and Minz, 2008).

References

Diao, Shuren 刁書仁, Yi, Xingguo 衣興國, 1994. *Jin sanbai nian Dongbei tudi kaifa shi* 近三百年東北土地開發史 (Development of Manchuria in the recent 300 years). Jilin wenshi Press.

Ding,Yizhuang 定宜庄 *et al.*, 2004. *Liaodong yimin zhong di qiren shehui* (遼東移民中的旗人社會 Immigration and Eight Banner Society in Liaodong). Shanghai kexue yuan Press.

Du Bois, C. and Tan, C-B, Mintz, S. eds., 2008. *The World of Soy.* National University of Singapore Press.

Fan, Jinmin 范金民, 2013. *Qingdai zhongqi Shanghai chengwei hangyun zhongxin zhi yuanyin tantao* (清代中期上海成為航運業中心之原因探討, Study on Shanghai becoming a maritime trade center in the mid-Qing), *Anhui shixue,* 1: 29–38.

Fu, Jia-chen, 2018. *The Other Milk: Reinventing Soy in Republican China.* University of Washington Press.

Huang, H.T., 2000. *Biology and Biological Technology. Part V: Fermentations and Food Science.* Vol. 6 of *Science and Civilization in China,* J. Needham ed., Cambridge University Press.

Isett, Christopher, 2007. *State, Peasant, and Merchant in Qing Manchuria, 1644–1862*. Stanford University Press.

Kato Shigeshi加藤繁, 2012. "Dongbei dadou doubing shengchan di youlai" (東北大豆豆餅生產的由來, Origins of soybean and soy meals produced in Manchuria). 1953 Trans. Wu Jie, in *Zhongguo jingji shi kaozheng* (中國經濟史考証Studies in Chinese economic history) v.2. Beijing: Zhonghua Press.

Kung, James and Nan Li, 2011. Commercialization as exogenous shocks: The effect of the soybean trade and migration in Manchurian villages, 1895–1934, *Explorations in Economic History* 48: 568–589.

Leung, Angela K., Becoming an everyday food: soy sauce in modern China (ca.1900–1930), in Fu, Jia-chen, King, Michelle, Klein, Jakob eds, *Modern Chinese Foodways.* forthcoming.

Leung, Angela K, 2019. To Build or to Transform Vegetarian China, *Moral Foods. The Construction of Nutrition and Health in Modern Asia,* Leung, A. and Caldwell, M. eds., Honolulu: University of Hawaii Press: 221–240.

Lian, Dajin 連大進, "Taiwan heidou: yongxu nongye mingri zhi xing" (台灣黑豆: 永續農業明日之星 Taiwanese black beans, rising star of sustainable agriculture). *Xiangjian xiaolu,* 1994.20.

South China Morning Post, China's soybean production to increase 40% by 2025 amid food security alarm, 14 January, 2022.

Wang, Jinling *et al.*王金陵等, 1999. *Zhongguo Dongbei Dadou* 中國東北大豆 (Soybean in Northeast China). Harbin: Heilongjiang kexue jishu Press.

Wang, Lianzheng王連錚, Guo, Qingyuan郭慶元, 2007. *Xiandai Zhongguo dadou* 現代中國大豆 (Contemporary Chinese soybeans). Beijing: Jindun Press.

Yi Baozhong衣保中, *Zhongguo Dongbei nongye shi*中國東北農業史 (Agricultural history of Chinese Manchuria). Jilin wenshi Press, 1993.

Zhao, Rongguang 趙榮光, 2005. "Zhongguo lishi shang di jiangyuan yu jiangyuna wenhua shulun — yi Shanghai shi di chuantong yu xinshi jiangyuan ziliao duibi wei zhu" (中國歷史上的醬園與醬園文化述論- 以上海市的傳統與新式醬園資料對比為主. A discussion on pickle shops and the culture of pickle shops in history mainly based on comparing the traditional and modern shops in Shanghai). *Yinshi wenhua yanjiu*, 4/16: 9–19.

https://doi.org/10.1142/9789811276736_0006

Chapter 6

Soybean Cultivation, Production and Quality Assurance

Shwu-Pyng Joanna Chen

State Key Laboratory of Agrobiotechnology,
The Chinese University of Hong Kong
joannaspt.chen@icloud.com

Soybean, as a valuable crop, production supply chain can be complex. Many elements can affect the production efficiency. Influencing factors impacting soybean production particularly seed variety and quality are explained. High quality seed is the root of successful cultivation expansion. International and national seed quality assurance programs and systems for ensuring high seed quality have been introduced. Since the African region is the fastest population growth area with food security and malnutrition concern, promoting soybean food value and providing farmers support programs for the area is encouraged.

A global seed market overview entailing the important role soybean plays in the world is outlined. Soybean as a valuable crop (crop of the century), its worldwide cultivation present actualities are briefly reviewed. Complex soybean value chain and related players internationally and nationally involved are introduced. Different aspects of soybean production from cultivation to quality assurance including varieties, research and development, registration/IP, testing, certification, labelling, farming technologies and commercialisation are discussed. Finally, the prospects of moving forward from today into future is provided. Supporting program links for Sub-Saharan Africa where future soybean production expansion is likely because of land availability and size are provided. The education information for recognising soybean food values is encouraged to use.

1. Seeds

According to UN population data and forecast (World population, https://www.worldometers.info accessed 19 April 2022), world population in 2020 is 7.8 billion and is projected to increase to 9.7 billion in 2050 and is expected to reach 10 billion in 2057. The projected highest growth area is Africa where 2/3 of world population growth will come from. Such fast increase of population outpaces increase in food supply. Food security becomes a global concern particularly in the African region.

Population growth increases agriculture food and feed demands. With the rapid bulging population, at least doubling current agriculture output is needed to feed the world. Good agriculture production starts with high quality seeds. Market size of global seed trade in 2021 is valued at US $63 billion and is projected to reach US $86.8 billion by 2026 recording a CAGR (Compound Annual Growth Rate) of 6.6% (Globe Newswire Report Linker 8 April 2021, https://www.reportlinker.com/p04436644/?utm_source=GNW accessed 19 April 2022). Seeds of major crops of food importance include maize (corn), soybean, rice, wheat, vegetables, and others.

Seeds are the basis of cultivation. They are critical assets for individual country's sustainable food supply. Several countries established their own germplasm seed banks to reserve these valuable resources. There are currently over 1700 seed banks worldwide. In 2008, Norway Svalbard Global Seed Vault was established to cover worldwide resources. This Vault stores 4.5 million different backup crop seeds currently and is set up to withstand major catastrophes to enable human agriculture continuation. (Svalbard Global Seed Vault, http://www.croptrust.org accessed 19 April 2022). According to Organization for Economic Co-Operation and Development (OECD) data, in 2021, there were 62,000 seed varieties covering 204 species registered in the worldwide seed trade (OECD, 2021). North America is the largest region followed by Asia Pacific in 2021 (Mordor Intelligence, 2022). Because of the importance of the seeds, many countries established their own seed laws and regulations to protect varieties, govern R&D, and fair trades. US seed law was promulgated in 1940 initially, Argentina 1973, Brazil 2003, and China 2000. These seed laws are regularly updated and enforced to reflect practical needs.

2. Soybean as a Valuable Crop

Cultivated soybean (*Glycine max* [L.] Merr.) belongs to Leguminosae or Fabaceae family. The same seed of soybean is used for planting/sowing, food/feed, and various applications. History of soybean dates back to China more than 4000 years ago. It was spread to Korea, Japan, EU, US, and Brazil at different time. Soybean offers high nutritional value. It contains high quality protein, healthy unsaturated fatty acids, low GI carbohydrate, high fibres and various bioactive compounds such as isoflavones. An example of soybean composition is shown in Figure 1. Typically it contains protein 38%, oil 18%, soluble carbohydrate 15%, insoluble carbohydrate 15%, water, ash and others 14%.

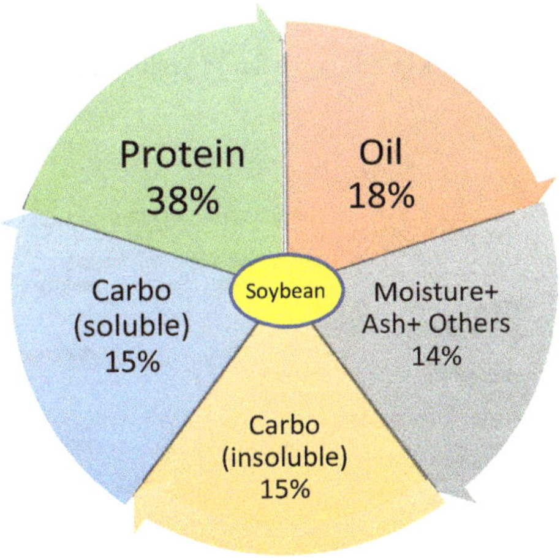

Figure 1. An example of soybean composition.

Utilisation of soybean worldwide includes uses as foods (~7%), oils (~13%), feeds/animal foods (~77%), and biodiesel (~2.8%). Soybean can be processed to yield meals and oils that offer high value applications (Figure 2). Roughly a soybean can produce 18% oil, 80% meal and remaining others. Soybean is unfortunately not well known as a human high nutritious food outside Asia particularly in Africa region. Reasons may be because of limited education and access to adequate food processing technologies. Soybean is unique as it is an excellent source of plant-based protein for food and feed. Due to its nitrogen fixation ability of the nodules, it is a beneficial crop for the environment. Soybean global planting and harvest schedule can be different in different areas and countries (Figure 3). North America, Asia Pacific, Eastern Europe plant during April to June (India July) and harvest in November/December. South America plants in November/December and harvest in March to July.

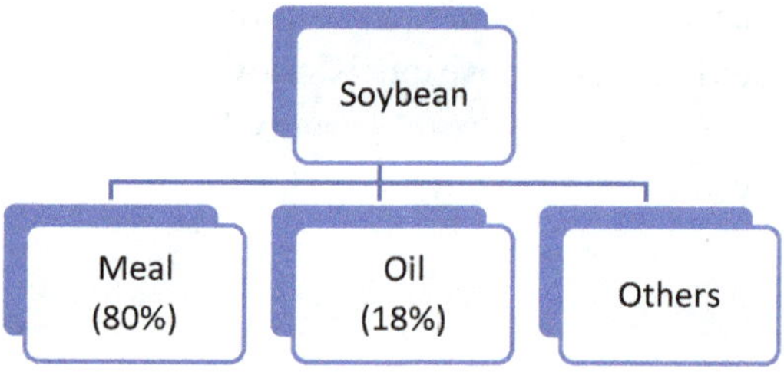

Figure 2. Processing soybean to yield meals and oils.

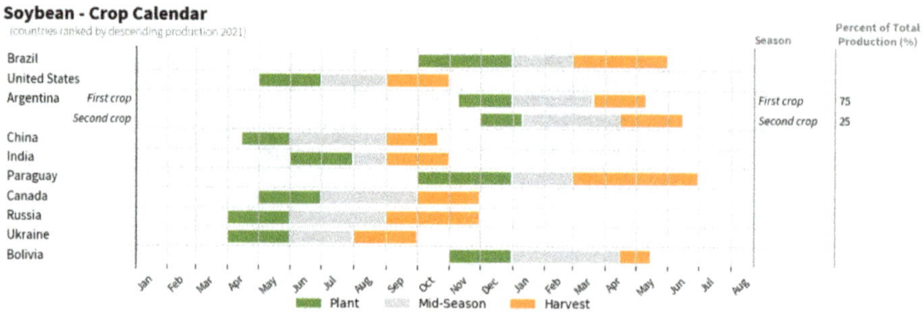

Figure 3. Soybean Global Planting and Harvest Schedule.

Source: USDA International Production Assessment Division www.ipad.fas.usda.gov

Some areas like Argentina offers second crop. Difference in harvest time offers flexibility on procurement in global trade.

3. Global Soybean Production

Globally, soybean production reached more than 363 million MT in 2021. Major production area, quantity and % contribution by country is shown in Figure 4. In 2021, Brazil produced the most (37% of world total) followed by the US (33%), Argentina (12%), China (5%) and India (3%). Top producing countries from 1961 to 2021 (Figure 5) are Brazil, USA, Argentina, China and India. In 2019, Brazil surpassed US as the top producing country due to a 25% tariff imposed by China on US soybeans in 2018. China is the largest soybean importing country accounting for over 60% of world exports followed by EU (<10%) and Mexico (<4%) (Import soybean worldwide 2021/2022: Statista, www.statista.com accessed 01 Feb 2022). Soybean production quantity in Brazil increases very fast in recent years (Figures 5, 6). Brazil started planting soybean in the South subtropical area and moved up to Midwest tropical area known as Cerrado where overcoming acidic poor-quality soil is required. Great efforts of Brazilian research

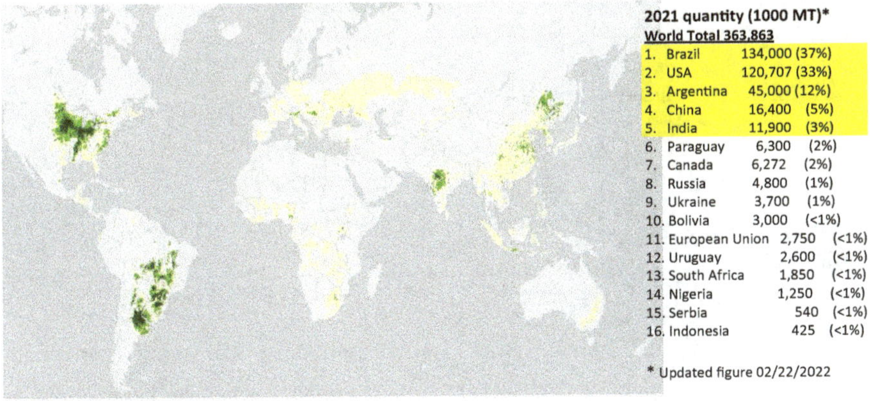

Figure 4. Global soybean production area and quantity 2021.

Source: USDA International Production Assessment Division www.ipad.fas.usda.gov accessed Feb 22, 2022

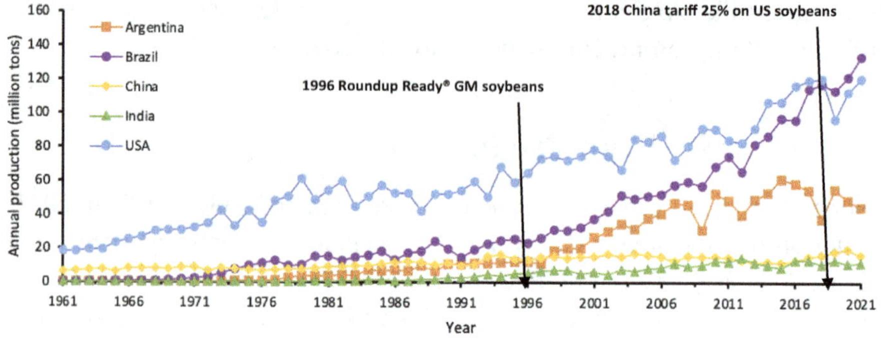

Figure 5. Soybean production in the top five producing countries 1961–2021.

Data source: 1961–2020; FAOstat (https://www.fao.org/faostat/en/) retrieved on 1 Feb, 2022. 2021; USDA (https://ipad.fas.usda.gov/cropexplorer/Default.aspx) retrieved on 24 Feb, 2022

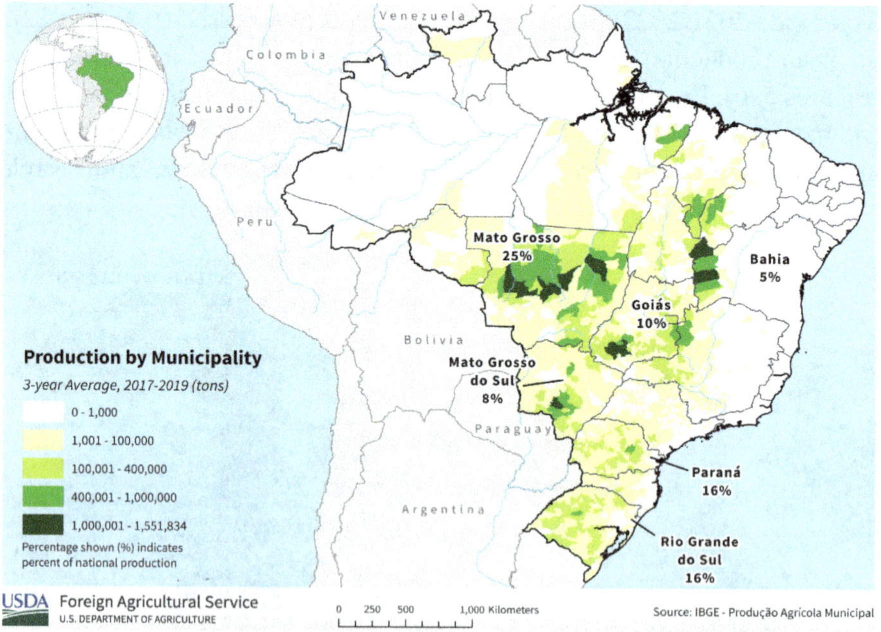

Figure 6. Brazil soybean production 3-year average (2017–2019) in tons.

USDA (https://ipad.fas.usda.gov/cropexplorer/Default.aspx) retrieved on 03/07/22

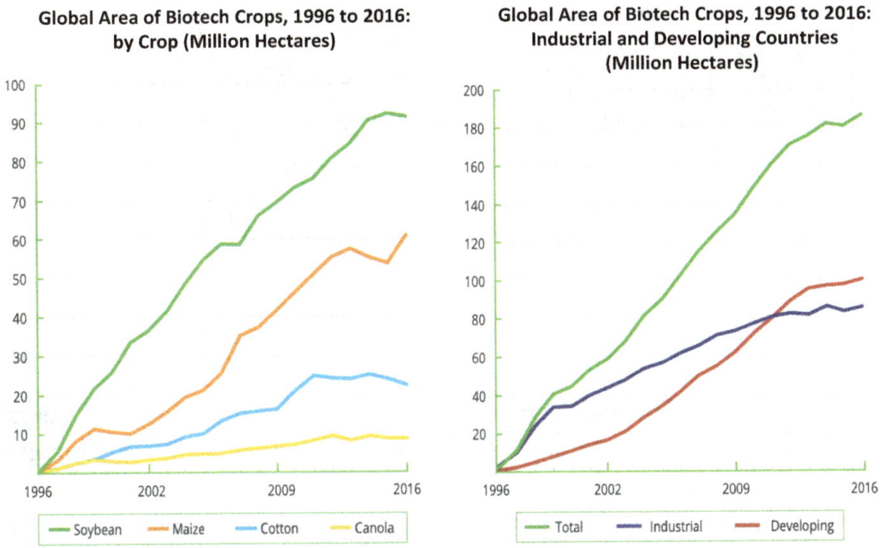

Figure 7. Global area of GM soybeans 1996–2016.

Clive James, ISAAA 2016 Global Status of commercial Biotech/GM crops, ISAAA brief No. 52, Figure 10. (L), Figure 1. (R), Ithaca, NY. ISAAA: International Service for the Acquision of Agri-Biotech Applications); FAOSTAT: 2016 Worldwide soybean total harvested area 122 million Hectares.

and development organization to generate new varieties suitable to grow in this region are highly commendable (Cattelan and Dall'sAgnol, 2018).

More than 80% of global soybean planted today are Genetically Modified (GM) varieties. More than 90% of the top three country output is GM soybeans (Figure 7). Significant quantity (>80%) of production globally is managed by large-scale farmers (Voora *et al.*, 2020).

4. Soybean Value Chain

Soybean value chain (Figure 8) involves various stakeholders. It starts with breeders who breed and generate different varieties. These varieties are then cultivated and produced by farmers. Farmers can use the produced seeds for sowing again or sell the seeds through traders for use in

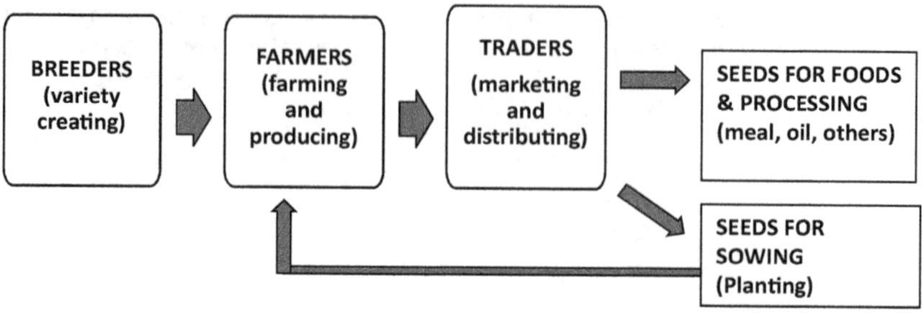

Figure 8. Soybean value chain.

either sowing or for foods and processing (meal, oil and others). Traders will market and distribute seeds for further commercialisation via domestic or international trades.

There are several international organizations who participate in setting/executing rules and regulations to govern the global seed trade movement and ensure seed ownership protection, quality guaranteed as well as trade fairness. These organizations include WTO, FAO, OECD, NDA, UPOV, AOSCA, ISTA, IPPC. These organizations inter-relationship and work flow is illustrated by OECD figure and is shown in Figure 9. WTO/FAO is the highest guiding principle in building capacity, providing aids and supervising trade agreements for variety development and use. Plant breeders and seed companies are responsible for variety maintenance. International trade movements are managed by ISF, OECD and phytosanitary by IPPC. OECD in collaboration with country NDA to ensure seed schemes are followed properly. The procedure is: plant breeders breed varieties; register and trial varieties, file protection to UPOV; OECD/NDA approve and list the varieties; seed traders/companies commercialise the varieties, OECD/AOSCA certify the seeds with ISTA, AOSCA methodologies.

5. Soybean Production

5.1. *Influencing factors*

There are many factors that impact the soybean production output. **Public policy and IP protection** is an important one which includes

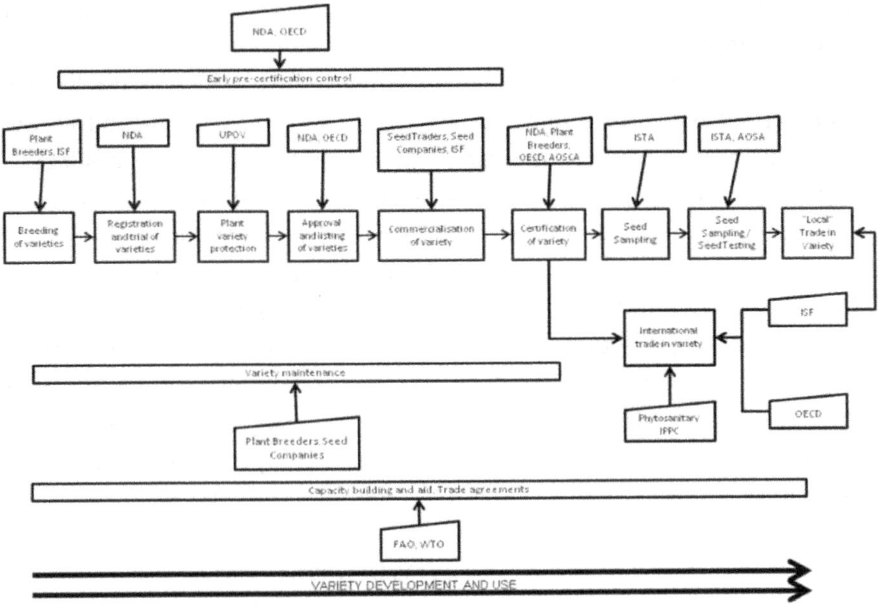

Figure 9. Overview of international organizations and other actors in the seed supply chain.

Source: OECD international seed trade schemes 2012, ANNEX1, Figure1, P14

OECD: Organization for Economic Co-operation and Development; **NDA**: National Designated Authority; **ISF**: International Seed Federation; **UPOV**: International Union for the Protection of New Varieties of Plants; **AOSCA**: Association of Official Seed Certifying Agencies; **ISTA**: International Seed Testing Association; **IPPC**: International Plant Protection Convention (prevent plant pest spread across international boundaries); **FAO**: Food and Agriculture Organization of the United Nations; **WTO**: World Trade Organization

country seed laws/rules/regulations execution/enforcement, farmers subsidy/tax/fee, violation penalty. It entails seed variety registration/listing and breeder's IP protection as well as fair trade protection. How the seeds should be tested with authorised organization/personnel with agreed methodology, qualified as a new variety to make the national/international list and certified/labelled to ensure trading quality is clearly defined. Since many organizations and players are involved in the whole value chain, a central designated organization to supervise and govern the end to end process is necessary. Education and training program to farmers on variety choice, farming/production skills/knowledge, seed trading, financial access is required at different stages to ensure success.

Country decision on which **crop to plant** (rotate) on which year affects output as there are rice, wheat, corn in addition to soybean to grow as food crops. Availability of country **arable land and water resources** affects agricultural production alongside of weather impacts such as drought, flood, fire, global warming/freezing. The other impacting factors include variety choice (elite variety that adapts best for the planting area), GM soybean adaption that can yield 1.5 times or higher with lower cost, public/private organization partnership_ on R&D, scale-up and education, investments devoted to farm mechanisation and technology modernisation and finally **help to small farm holders** to enable accessing quality seeds, cultivation/production knowledge and financial operation for ultimate profitability.

A. Soybean variety, research and development

Soybean exists in wild for many years. Natural evolution to adapt growing conditions and human selections for cultivating easier and producing more has resulted in many different varieties. These soybeans exhibit different physical profiles such as different skin colour, stand-up versus lateral growth and different nutrient composition (Han *et al.,* 2021).

Conventional breeding has been used as techniques for variety selection until genetic modification technology appears. Genetic modifications create many new varieties with different traits since 1996 Roundup Ready™ which offers only herbicide tolerance initially. Today, three generations of GM soybeans exist including soybean characters in herbicide tolerance (HT), insect resistance (IR), combating climate change, improving nutrition and quality. The first generation offers single trait (HT), double stacked (HT and IR) and multiple stacked (herbicide, insect, weed and pest). Second generation offers soybeans that are tolerant to drought, salt and cold. The third generation offers nutrient enhancement such as higher oleic acid and omega-3-fatty acid (ISAAA, 2016).

Consumers and Green movement organizations resist GM soybeans fearing GM can cause long term harm to human and animal health. Since the introduction of the GM soybeans (more than 20 years), there has not been a direct adverse case linked to authorised GM foods. EU, the most strict area, declares safety on pre-authorized GM foods and feeds in 2015 to allow flexibility in consumption and planting with legislation change. Genetic modification is a process to alter selected genes benefiting desired profile in growth. Natural evolution and conventional breeding also goes through gene alteration/mutation process except the natural and conventional breeding way takes much longer time with less control. Nevertheless, to introduce any GM variety to any country/area, country regulatory approval is required to conduct a complete set of biosafety risk assessment submission.

B. Soybean variety IP protection

Number of soybean varieties owned by each country differs by country. A very large number exists. China germplasm bank collects 43,000 kinds. Every year, more varieties are created. In 2020, China alone approved 3112 (national 491 and local 2621). Internationally, soybean variety registration and list is handled by (OECD Schemes-varieties www.oecd/org/agriculture/seeds/varieties/). At national level, US (Agricultural Marketing Service), China (Ministry of Science & Technology Crop Germplasm information Platform; National new variety publication by Ministry of Agriculture and Rural Affairs) and Brazil (National registration of seeds & Seedling, National Registration of cultivars) each has her own registration process and registered list.

For IP protection, International organization is UPOV (DUS Distinctness, Uniformity, Stability Rule, TPG/9-11). The PLUTO plant variety database contains information on plant varieties from UPOV and OECD members. This database is a similarity search tool. It is available for public use with a fee (Pluto Plant Variety Database, https://pluto.upov.int). Nationally, US (Plant Variety Protection Act., Plant Patents, Utility Patents), China (Seed Law, State Council Order #213 Protection of new varieties) and Brazil (Plant Variety Protection Law No. 9.456) each has her own venues to handle variety and breeder IP protection. For multinational company who has large operation globally is better advised to protect their variety via IP registration. A 2019 legal case was raised by PepsiCo suing Indian farmers for planting their protected Lays potato chip variety is a good example. Without good protection, a good variety may be planted out of control.

C. Testing

a. Testing (Sowing Seeds)

Sowing soybean seeds are tested by following international rules specified by ISTA (International Seed Testing Association) and/or US AOSA (Association of Official Seed Analysts). Nationally authorised/approved laboratories and technicians are qualified to conduct testing and prepare reports. A sowing seed testing laboratory generally can handle sampling and conduct assorted testing such as purity, germination and others that are listed in Table 1. ISTA and AOSA rules are similar. China GB/T3543

Table 1. Soybean seed testing rules (sowing seeds).

No.	**Test item**	**ISTA rules**	**AOSA rules**	**GB/T 3543 rules (based on ISTA 1993)**
	Year established	1924	1908	1995 update
1	Sampling	✓	✓	✓
2	Purity	**Pure seed, other seed, inner matter**	**Pure seed, other crop seed, inner matter, weed seed**	**Pure seed, other seed, inner matter**
3	Germination	✓	✓	✓
4	Species and variety	✓	✓	✓
5	Moisture content	✓	✓	✓
6	Tetrazolium	✓	✓	✓
7	Seed health	✓	✓	✓
8	Thousand-seed Weight	✓	✓	✓
9	Coated seeds	✓	✓	✓
10	Excised embryo test for viability	✓	✓	X
11	X-ray test	✓	✓	X
12	Seed vigour testing	✓	✓	X
13	AP (Adventitious Presence) and GMO	✓	✓	✓

rules are based on ISTA 1993's and differ slightly on three items, excised embryo test for viability, X-ray test and seed vigour testing. Each country/state/province may have their own standards to comply. Table 2 lists examples of standard used in US Iowa State and China National standard. Seeds are classified into different classes such as foundation/registered/classified in US and basic/qualified in China with respective standards specified on purity, germination rate. The US specifies weed and noxious

Table 2. Sowing seed quality standard examples.

US Iowa Seed Certification Requirements

Soybeans

Quality Goals	**Foundation**	**Registered**	**Certified**
Pure seed (min.)	99.0%	99.0%	99.0%
Total inert matter	1.0%	1.0%	1.0%
Other crop seed (max. per lb.)			
Corn and sunflower	None	None	1per 5 lbs.
Additional (excluding forage crops)	2	2	2
Common weed seed (max. per lb.)	1	1	2
Secondary noxious weed seed (max. per lb.)	None	None	1per 5 lbs.
Primary noxious weed seed (max. per lb.)	None	None	None
Germinat on (min.)	80%	80%	80%
Minimum germinat on before penalty	70%	70%	70%

Source: VII.C Quality goals and Min Standards/Soybeans page 6, Iowa Seed Certification Regulations 2020, http://www.iowacrop.org/pdf/Approved_Conditioner_Handbook

China GB 4404.2-2010 Seed of Food Crops Part 2: Legume

Seed Classification — Soybean %

Crop variety	**Seed classification**	**Varietal purity**	**Cleanness (pure seeds)**	**Germination rate**	**Moisture**
Soybean	Basics seeds	≥99.9	≥99.0	≥85	≤12.0
	Qualified seeds	≥98.0			

Note: The moisture of soybean seeds north of the Great Wall and in alpine regions is allowed to be higher than 12.0% but not higher than 13.5%. The moisture of soybean seeds south of the Great Wall (excluding alpine regions) is not allowed to be higher than 12.0%.

Source: http://www.gbstandards.org/

seed amount but not in China. China specifies moisture content but not the US.

b. Testing (seeds for Foods and Processed — oil, flour, concentrate, isolate, other specialities)

Soybean seeds intending for direct food consumption or industrial processes to make soybean oil for cooking/biodiesel, meals for plant-based human food formulation and feeds as protein source to pig, poultry, cattle, fish can be tested following food and food ingredient analysis methods. Methods like AOAC (Association of Official Analytical Chemists) to test fat, protein, and other compositions. Countries also specify their own testing methods in line with official methods, e.g. China GB5009.5-2016 for Fat test and GB 5009.6-2016 Protein. The laboratory set up is the same as that of normal food and ingredient testing. Lab accreditation and technician authorisation/certification is required to ensure authenticity of the results. Testing items can cover physical ones like size, colour, impurities, appearance, and chemical ones (moisture, protein/amino acids, fat/fatty acids, fibre, GMO and others. Additional microbiological testing (bacteria, yeast, mold count and mycotoxins) is required because of safety concern. Standard example used to qualify the goods in China are shown in Table 3. In general, seeds are graded based on % sound kernels, % damaged kernels, % impurities, % moisture, colour, and odour.

D. Sowing seeds certification, labelling and compliances

Purpose of certifying seeds intended for sowing is to preserve genetic purity and varietal identify to ensure seed of high quality. The process is defined clearly in the OECD seed scheme for globally traded seeds. The scheme includes several activities such as land requirement, planting eligible stock, field inspections, meeting standards and proper labelling. The stepwise process is outlined in OECD 2021 certification process seed scheme flyer Figure 10. It starts with sample submission followed by crop sown and inspected while grown on the field. Seeds are then harvested, processed according to proper protocols. Seed sample is then sent for laboratory testing. Once all tests are done and requirements met, seed lots are then labelled and OECD certification is issued. Country specific certification scheme is aligned with the OECD scheme via National

Table 3. Soybean quality standard (China GB 1352–2021 replacing 1352–2009).

Table a Soya Bean Quality Indicators

Grade	Rate of sound kernels/%	Rate of damaged kernels/%		Impurities content/%	Moisture content/%	Colour and odour
		Total	Rate of heat-			
1	≥95.0	≤4.0	≤0.2	≤1.0	≤13.0	Normal
2	≥90.0	≤6.0	≤0.2			
3	≥85.0	≤8.0	≤0.5			
4	≥80.0	≤10.0	≤1.0			
5	≥75.0	≤12.0	≤3.0			
Off-grade	<75.0	—	—			

Note: "—" indicates that there is no requirement.

Table b Quality Indicators of High-oil Soya Bean

Grade	Fat content/%	Rate of Sound	Rate of damaged kernels/%		Impurities content/%	Moisture content/%	Colour and odour
			Total	Rate of heat-			
1	≥22.0	≥85.0	≤8.0	≤0.5	≤1.0	≤13.0	Normal
2	≥21.0						
3	≥20.0						

Table c Quality Indicators of High-protein Soya Bean

Grade	Protein content/%	Rate of sound	Rate of damaged		Impurities content/%	Moisture content/%	Colour and odour
			Total	Rate of heat-			
1	≥44.0	≥85.0	≤8.0	≤0.5	≤1.0	≤13.0	Normal
2	≥42.0						
3	≥40.0						

Source: http://std.samr.gov.cn/search/std?q=GB%201352&tid=gb

Designated Authority. Besides OECD, other certification organizations include ISTA and AOSCA (Association of Official Seed Certifying Agencies). The detailed OECD soybean scheme is outlined in Table 4 covering general requirement, procedure, label and container marking,

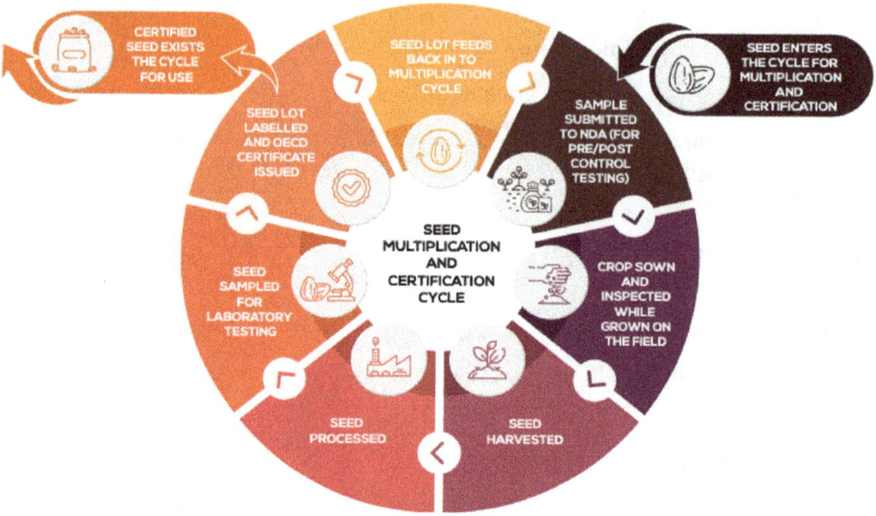

Figure 10. OECD Certification Process, Seed Scheme Flyer 2021.

NDA: National Designated Authority

Table 4. OECD seed scheme ruiles and regulations 2021.
Grass and Legume Seed — *Glycine max* (L) Merr. — Soybean

General

- Must be species listed eligible for certification
- Apply common basic principles & method of preparation
- Apply rules and regulation of the relevant scheme
- Ensure implementation by NDA

Procedures

- OECD list of variety
- Seed category (Pre-basic, Basic, Certified)
- Production of seeds
- Control of seed production (requirements & field inspection)
- Seed lot sampling (max lot size 30,000 kg) and sample storage
- Seed analysis
- Issue certificate per seed category filed
- Post control tests of seeds (procedure & frequency)
- Seed lots (homogeneity, size and fastening containers)
- Container ID and others (reference numbers for certificates and seed lots)

Table 4. (*Continued*)

OECD Label and Marking seed containers (common)
• La colour
- Pre-basic seed => white with diagonal violet stripe
- Basic seed => white
- Certified seed, 1st generation => Blue
- Certified seed, 2nd generation or more => Red
- Not finally certified seed => Grey
• Reference to the OECD scheme:
OECD Seed Scheme on label or outside the container
• Label information:
- Name & address of NDA
- Species (Latin name)
- Variety denomination (or synonym)
- Category (pre-basic, basic, certified, generation #)
- Lot reference number
- Declared net/gross weight
- Serial number identifying each label
- Region or country of production
- Statement of re-packing and re-labelling
- Others
OECD Soybean Seed Scheme specifics (under Grass and Legume seed)
• Previous cropping (min 3 years)
• Isolation (avoid contamination)
• Weeds (should not contain excessive number)
• Number of harvest years
• Field inspection (at flowering time for legumes)
• Varietal purity (Glycine max)
Basic seed 99.5%
Certified seed 1st generation: 99.0%
Certified Seed 2nd generation: 99.0%

and specifics on soybean seeds standards to be met. In the US, the OECD scheme is carried out by AOSCA and State Seed certifying Agency under a corporate agreement with the USDA Agricultural Marketing Service (AMS).

The required information and colour coding put on label upon completing certification can be demonstrated using US Maryland seed label (Table 5, Example 1). China does not follow OECD labelling but requires label contents to include information such as % variety purity, % purity seed, % germination rate and % moisture as shown in Table 5, Example 2.

Additional certifications and compliances are often requested to meet by different organizations and users to show natural and sustainable soybean status. These include Organic/nonGMO, ProTerra (by ProTerra Foundation), RTRS (Round Table on Responsible Soy Association), CRS

Table 5. Seed label examples (required information and colour coding).

Example 1: US Maryland seed label (the label contents can vary among states)

Reliable Seed Company
123 Production Way
Seedville, VA 12345 USA
Phone: (123) 456-7890

'Rumsey' yellow indiangrass
Sorghastrum nutans

NET WEIGHT:	25 lbs	PURE SEED:	93.80%
LOT #:	IB097	OTHER CROP:	0.00%
DATE TESTED:	January 2006	INERT MATTER:	5.60%
ORIGIN:	PA	WEED SEED:	0.60%
		GERMINATION:	32.00%
NOXIOUS:	21/lb Giant Foxtail	HARD SEED:	0.00%
	15/lb Wild Garlic	DORMENT:	61.00%

And a second label indicating the certified class of seed: example BLUE for class of CERTIFIED SEEDS

Englert, J.M. 2007. A Simplified Guide to Understanding Seed Labels. Maryland Plant Materials Technical Note No. 2. USDA-NRCS National Plant Materials Centre, Beltsville, MD. 3p.

Example 2: China seed label

Varietal Purity	Purity seed	Germination	Moisture content
≥98.0%	≥98.0%	≥85%	≤12.0%
Net weight	Species	Category	
10kg	Soybean	Qualified seed	

Business Permit No: D(S) Business License of Crop Seeds (2016) No.1 Quality standards: GB 4404.2-1996

(Cefetra Certified Soya), Rainforest Alliance, ISCC (International Sustainability and Carbon Certification), and VSS-compliant (Voluntary Sustainability Standards*). South American countries are also requested to sign Amazon Soy Moratorium 2006 and Cerrado Manifesto 2017 to commit for curbing deforestation and loss of natural habitats. All these make soybean production process and supply chain more complex (Voora *et al.*, 2020).

E. Farming technology

To achieve high yield and quality soybean production, a series of farming techniques must be well versed by farmers. These include 1. **Soil quality checking** to ensure soil is well drained, pH > 6.5, complete/balanced mineral contents, and quality bacteria population. 2. **Nutrients application** for inoculating beneficial bacteria and adequate fertiliser application. 3. **Weather and irrigation planning** to make sure weather data can be monitored/tracked for anticipating rain fall, drought, storms, temperature, flood so proper irrigation can be planned. 4. **Weed, pest and disease management** so adequate amount of herbicide, pesticide can be applied and disease (e.g. Rust) can be prevented. 5. **Production management** for right variety selection, seeding rate decision, planting timing, crop growth monitoring, harvest/post-harvest loss reduction, cleaning, grading, drying, bagging, warehousing, storing done properly followed by sampling/certifying and labelling for trade. 6. **Farm mechanisation** where possible to minimise labour use and loss reduction. Digital decision support tools such as high resolution soil map, Real-time crop monitoring and Bayer Climate Fieldview™ are available commercially to help farmers improve skills and techniques. Investing in modern farming technology can increase efficiency greatly (Varshney *et al.*, 2021).

---

*VSS: UN definition is "standards specifying requirements that producers, traders, manufacturers, retailers or service providers may be asked to meet, relating to a wide range of sustainability metrics, including respect for basic human rights, worker health and safety, the environmental impacts of production, community relations, land use planning and others.

F. Seed commercialisation

Successful seed commercialisation involves promoting elite variety to farmers, training and educating commercial processes and handling knowledge, and providing financial support and sponsorship where applicable. The variety promotion is from breeders to farmers to allow access and select the most suitable seeds for their land cultivation. Institutional support and digital agriculture tools introduction helps maximise yield. The training and education should include front line field demonstrations, organised study groups/classes, posters, newsletters and in-person communications to be effective. Farmers Associations can support to market and trade seeds by giving farmers access to electronic trading portals such as **Wefarm** https://about.wefarm.com/, **eNAM** https://www.enam.gov.in/web/ **and Digital Mandi** https://www.iitk.ac.in/MLAsia/digimandi.htm.

6. Going Forward to Produce More

While growing and producing more, several areas should be specially tended to maximise output and achieve long term sustainable supply. These include **sustainable production**, **GM technology maximisation**, **modern farming technology application and accessibility**, **farmers' education and training** and **financial support** such as subsidy and accessible funds for small farm holders.

On sustainable production, global publicised issues such as deforestation, climate change (+2^{0}C in 2050), biodiversity loss, soil quality erosion, herbicide resistance, wild weeds increasing, labor right violation need to be addressed. FAO advocated Conservation Agriculture System* and digital agriculture should be promoted. Complying to different sustainable initiatives are encouraged along with origin tracing and supply chain assurances.

---

*Conservation Agriculture System is a farming system aiming to reduce soil degradation and improve sustainable crop production and environmental quality. Practices include crop diversification (intercropping/crop rotation or both), minimum soil disturbance (no tilt), and permanent soil organic cover (source: FAO 2018 Promoting Conservation Agriculture in Timor-Leste http://www.fao.org/publications).

GM technology can create elite varieties for high yield and temperature/drought tolerance, weed killer/insecticide/disease resistance, higher nutrition. GM technology also shortens breeding time to enable faster production. It should be promoted since safety has not been an issue. Modern farming technologies and tools need to be made assessable to farmers particularly small farm holders. Education and training program particularly in developing countries to ensure complete soybean supply chain management knowledge should be emphasised. It is also critical to educate population in rural underdeveloped areas the value of soybean as a food crop and various processing methods to turn soybean into nutritious foods. Finally critical financial support to small farm holders is essential. Only profitable farmers are happy to grow more.

7. Promote Soybean Food Value

African staple is Maize. Maize based diet is generally lack of protein quality and quantity. Maize protein is deficient in essential amino acid such as lysine and tryptophan while soy protein is considered complete with 9 essential amino acids. Protein content of cooked corn (Maize) kernels without cob is 3.2% while soybean dry mature seed is 34.1% (Adams, 1975). Maize based diet without supplements can cause malnutrition easily (FAO, 1992). Soybean as a crop is relatively unknown to African rural households (Dlamini *et al.*, 2014). If the diet can combine maize and soy, the nutritional value will be greatly increased.

Soybean with its high nutritional value is versatile. It can be consumed fresh as edamame, bean sprout, roasted beans. It can be made into soymilk (plain or with assorted flavours like strawberry, chocolate, vanilla), tofu, tofu skins. Tofu can be consumed in different ways such as cheese replacement, mix as salad dips, roasted cubes to mix or stir fry with vegetables. It can be fermented to make Natto (a famous Japanese food), soy sauce, different types of sauces (Jiang in Chinese and Korean), and tempeh (a local Indonesian nutritious food). The soy protein fractions can be used as protein bases to make plant-based meats such as beef burgers without beef, chicken nuggets without chicken and imitation seafoods without real seafoods. These great soybean food value must be promoted heavily to needy areas to end malnutrition and achieve food nutrition security in the long term.

8. Africa Support Programs

African region has the fastest population growth and needs much help to improve agriculture knowledges and practices in different fronts as well as educate soybean food value particularly to those small holder farming families in poor rural Africa. Many programs are in place to carry out such tasks. Below is a list of a few that are active.

FANRPAN (The Food, Agriculture and Natural Resources Policy Analysis Network): Policies for a food secure Africa. https://www.fanrpan.org

ATONU (Agriculture to Nutrition — Farm Africa, Funded by Bill and Melinda Gates Foundation): Provide training & knowledge on nutrition, hygiene, homestead-based vegetable production. 33% of childhood deaths in sub-Saharan Africa are linked to under-nutrition. https://www.fanrpan.org/projects/ATONU

BAYER Farmer Development Program of South Africa (partner with GSA Grain South Africa). Develop small-holder and emerging farmers one step at a time. https://www.cropscience.bayer.africa

There are also many NGO and volunteer groups working in African countries to help improve their livelihoods. It is most advisable to work in harmony with major organizations to maximise the result.

References

Adams Catherine F., 1975. Nutritive Value of American Foods in common Units. USDA Agriculture Handbook No. 456. pp. 67 and 155. United States Department of Agriculture (USDA) Agricultural Research Service.

Cattelan, A.J., and Dall'Agnol, A., 2018. The rapid soybean growth in Brazil. *OCL,* 25(1): D102, 1–12. https://doi.org/10.1051/ocl/2017058

Dlamini Thula Sizwe, Tshabalala Precious and Mutengwa Tafadzwa, 2014. Soybean production in South Africa. *OCL*, 21(2): D207. DOI: 10.1051/ocl/201347

FAO (Food and Agriculture Organization) of UN (United Nations), 1992. Maize in human nutrition, Chapter 8. In: *Improvement of Maize Diets*, Rome, Italy. FAO Food and Nutrition Series, No. 25, ISBN 92-5-103013-8.

Han, T.F., Zhou, X., Guan, R., Shu, S., Tian, S., Wang, S., and Yang, Z. Yesterday, today and tomorrow of soybean seed industry (in Chinese). Available online: http://www.zys.moa.gov.cn/mhsh/202105/t20210513_6367666.htm (accessed on 01 January 2022).

ISAAA. Global Status of Commercialized Biotech/GM Crops, 2016. ISAAA Brief No. 52. 2016. (ISAAA: International Service for the Acquisition of Agri-Biotech Applications).

Mordor Intelligence, 2022. Seed market-growth, Trends, CORVID-19 impact, and forecast (2022–2027) Home>Industry reports>Agriculture>Seed Market, accessed 072122. https://www.mordorintelligence.com

OECD, Organization for Economic Cooperation and Development, 2021. *OECD Schemes for the Varietal Certification or the Control of Seed Moving in International Trade*; OECD: Paris, France, 2021.

Varshney R.K. *et al.*, 2021. Rapid Delivery Systems for Future Food Security, *Nat. Biotechnol.* 39, 1179–1181. https://doi.org/10.1038/s41587-021-01079-z

Voora V. *et al.*, 2020. Global Market Report: Soybeans, Sustainable Commodities Marketplace Series Oct 2020, IISD.

https://doi.org/10.1142/9789811276736_0007

Chapter 7

Phenotyping as an Indicator of Genetic Diversity in Soybean

Ndiko Ludidi

Department of Biotechnology,
University of the Western Cape Bellville, South Africa
nludidi@uwc.ac.za

1. Introduction

Soybean (*Glycine max*) is a protein and oil rich crop mainly cultivated for livestock feed, especially for beef and chicken. Oil extracted from soybean seeds is used as cooking oil and as a component of other food products such as margarine, chocolate and ice cream. The oil is also used in the manufacture of cosmetics or soaps. Soybean is used as food in many East Asian countries, where it is consumed as soymilk, tofu, soy sauce and edamame, among many. Soybean oil has also become a major source of raw material for biodiesel production. These many uses of soybean have resulted in the production of soybean increasing by more than double over the last two decades. This has been enabled by creation of new varieties with desirable agronomic traits and adaptation to various agroclimatic zones.

Soybean cultivation places a significant demand on water resources, with an average of 650 litres of water required to produce 1 kg of soybean dry matter. Therefore, selection of soybean lines with good water use efficiency is of great importance. Variation in yield, water use efficiency and adaptation to agroclimatic zones are driven by genetic diversity. In this context, genetic diversity is the range of different inherited traits within a species. Hence, high genetic diversity implies that many accessions/lines will have a wide variety of different traits/phenotypes. It thus follows that the extent to which phenotypes vary for a specific trait reflects genetic diversity within a species. This makes phenotyping an important indicator of genetic diversity. Plant phenotyping assesses complex plant traits in order to describe the anatomical, ontological, physiological, and biochemical properties of a plant.

2. Relationship Between Genotypic and Phenotypic Diversity

The genotype of an organism reflects the complete set of genetic material of the organism, which gives rise to alleles — i.e. two or more of alternatives of a gene that contribute to the determination of a phenotype — a combination of whose dominant, heterozygous or recessive nature influence the phenotype of the organism. Variation in the genotype of the of individuals within a species (i.e. genetic diversity) will thus lead to variation in the phenotype of individuals in the species (i.e. phenotypic diversity). Genetic diversity can be measured using the genetic coefficient of variation, whereas phenotypic diversity can be measured based on the phenotypic coefficient of variation.

On the one hand, the genetic coefficient of variation refers to a mean-standardised index that reflects the evolution of a trait within a population in a species (Hill, 2010; Houle, 1992), and thus the potential of a particular trait to respond to natural selection in a population. On the other hand, the phenotypic coefficient of variation reflects the combined contribution of the genotypic variance and variance in the environment within which a population of a species exists. It thus follows that expression of a phenotype within a population will vary as a result of not only the genotype but also the environment. It is important to note that expression of a phenotype is also dependent on interactions between genetic loci, and thus such epistatic interactions will influence phenotypic diversity. Epistasis refers to an interaction of genetic variation at two or more loci in which the phenotype is not simply the additive combination of effects determined by individual loci. This implies that many phenotypes are a result of interactions between various loci and this can contribute to phenotypic diversity. The particular source phenotypic variation determines if a trait is influenced by natural or artificial selection and if the trait is influenced by environmental changes.

In consideration of the above concepts, it is crucial to appreciate the relative importance of both the genetic factors and the environmental factors in determining a particular phenotype. Heritability (i.e. the extent to which differences in genotypes account for different phenotypes) will lead to closely related individuals having similar phenotypic features. Consequently, genetic diversity is a determinant of differences in the phenotype of individuals within a population of a particular species. Given that determination of the genetic makeup of an individual requires sophisticated molecular genetic analysis, which is not easily accessible in resource-limited environments, phenotyping can be used as a means to infer genetic diversity.

3. Morphological Assessment and Genetic Diversity

Phenotyping based on morphological characteristics is useful in the selection of soybean varieties suitable for some agronomic traits. Morphological characteristics such as shoot height, leaf number, leaf shape, shoot dry weight, root length and root dry weight are influenced by interaction between the genotype and the environment, indicating a role of genetic diversity in determining morphological diversity. This implies that differences in soybean morphology/architecture can be used as a tool to infer genetic diversity. A good example of the use of morphological/architectural traits is root phenotyping work in genetically diverse soybean lines (Liu *et al.*, 2021). Root traits are important determinants of soybean capacity to access water and nutrients (Araki *et al.*, 2002; Rincon *et al.*, 2003). This becomes very relevant for drought tolerance because plants with long roots and large volume can penetrate deeper into the soil to access moisture in deeper soil depths since shallow soil depths would be depleted of soil under reduced precipitation (Araki *et al.*, 2002; Rincon *et al.*, 2003).

Equally important is the better ability of roots with large biomass, which are correctly angled, to acquire nutrients in the root zone of the soil — which is highly relevant for efficient nutrient uptake in nutrient-poor soils, especially soils deficient in phosphorous (Vandamme *et al.*, 2013). Furthermore, root architecture has significant impact on shoot traits, as shown by superior shoot dry weights in soybean genotypes that possess extensive root systems compared to small root systems (Salim *et al.*, 2021). This root-shoot relationship is presumed to be a consequence of increased leaf photosynthesis driving high carbon supply to the roots to promote root biomass to the benefit of the plant (Salim *et al.*, 2021). Besides root architecture, shoot architecture bears relevance as a tool for assessing genetic diversity. In this regard, leaf number, number of branches, leaf shape and leaf size are potential phenotyping traits that can be used to assign genetic diversity. These shoot traits can influence physiological parameters such as photosynthetic efficiency, transpiration and thus the extent of water loss from the plant canopy.

4. Molecular Biology as a Tool to Harness Benefits from Genetic Diversity

Despite their usefulness in providing basic indication of genetic diversity, morphological or architectural traits can be labour intensive and lack accuracy, have low throughput and limited detection power that are otherwise required to attain full benefit for unequivocal assignment/detection of genetic diversity. For this reason, molecular biology is highly useful to overcome the limitations of morphological traits in detecting genetic diversity. Various approaches for detecting genetic diversity include restriction fragment length polymorphism (RFLP), random amplified polymorphic DNA (RAPD), amplified fragment length polymorphism (AFLP), simple sequence repeats (SSR), single nucleotide polymorphism (SNP) and diversity arrays technology (DArT). Of these technologies, SNP and DArT approaches have proven to be very useful and robust in detecting genetic diversity, thus playing a pivotal role in associating genetic diversity with phenotypes and plant responses to their environments. These efficient genetic diversity detection tools are critical for genetic improvement of soybean in relation to yield and adaptation to environmental stress.

With the use of DArT, DNA polymorphisms identified in genetically diverse lines of crop plant species can be linked to specific traits by comparing the detected polymorphisms among the diverse lines to the differences in the phenotypes of the lines. Such phenotypes include yield, abiotic and biotic stress responses and various agronomic traits. Examples of the use of DArT in soybean include the application of DArT in developing genotypic data related to soybean yield components (Czembor *et al.*, 2021) and characterisation of soybean genetic structure and diversity in combination with SNP data (Shaibu *et al.*, 2022). Nutritional content, specifically protein content, has been associated with genetic diversity in populations of soybean, for which the genetic analysis for such diversity was demonstrated using a combination of SNP, SSR, DArTseq and DArT markers (Samanfar *et al.*, 2019).

Quantitative trait loci (QTL) analysis for analysis of responses of soybean to drought has been achieved by integrating the use of DArT markers in the analyses (Vu *et al.*, 2015). Various approaches for integrating phenotyping, conventional breeding and molecular markers in improving soybean tolerance to heat have been described (Jianing *et al.*, 2022) and hold promise for the use of such approach in improving soybean tolerance to other abiotic and biotic stresses. Future efforts towards achieving soybean adaptation to the effects of climate change (e.g. increased disease pressure, drought, flooding and heat stress) will be accelerated if integrated approaches such as those described in Jianing *et al.* (2022) are adopted. The speed with which such adaptation will be achieved will be enhanced if omics technologies such as transcriptome-wide RNA Sequencing, proteome-wide proteomic profiling and metabolomics constitute these endeavours based on analyses on soybean populations of diverse genetic backgrounds.

References

Araki, H., Morita, S., Tatsumi, J., and Iijima, M. 2002. Physio-morphological analysis on axile root growth in upland rice. *Plant Prod. Sci.*, 5: 286–293.

Czembor, E., Czembor, J.H., Suchecki, R., and Watson-Haig, N.S., 2021. DArT-based evaluation of soybean germplasm from Polish Gene Bank. *BMC Res. Notes*, 14: 343.

Hill, W.G., 2010. Understanding and using quantitative genetic variation. *Phil. Trans. Royal Soc. B*, 365: 73–85.

Houle, D., 1992. Comparing evolvability and variability of quantitative traits. *Genetics*, 130: 195–204.

Jianing, G., Yuhong, G., Yijun, G, Rasheed, A., Qian, Z., Zhiming, X., Mahmood, A., Shuheng, Z., Zhuo, Z., Zhuo, Z., Xiaoxue, W., and Jian, W., 2022. Improvement of heat stress tolerance in soybean (*Glycine max* L), by using conventional and molecular tools. *Front. Plant Sci.*, 13: 993189.

Liu, S., Begum, N., An, T., Zhao, T., Xu, B., Zhang, S., Deng, X., Lam, H.-M., Nguyen, H.T., Siddique, K.H.M., and Chen, Y., 2021. Characterization of root system architecture traits in diverse soybean genotypes using a semi-hydroponic system. *Plants*, 10: 2781.

Rincon, C.A., Raper, C.D.J., and Patterson, R.P., 2003. Genotypic differences in root anatomy affecting water movement through roots of soybean. *Int. J. Plant Sci.*, 164: 543–551.

Salim, M., Chen, Y., Ye, H., Nguyen, H.T., Solaiman, Z.M., and Siddique, K.H.M., 2022. Screening of soybean genotypes based on root morphology and shoot traits using the semi-hydroponic phenotyping platform and Rhizobox technique. *Agronomy*, 12: 56.

Samanfar, B., Cober, E.R., Charette, M., Tan. L.H., Bekele, W.A., Morrison M.J., Kilian, A., Belzile, F., and Molnar S.J., 2019. Genetic analysis of high protein content in 'AC Proteus' related soybean populations using SSR, SNP, DArT and DArTseq markers. *Sci. Rep.*, 9: 19657.

Shaibu, A.S., Ibrahim, H., Miko, Z.L., Mohammed, I.B., Mohammed, S.G., Yusuf, H.L., Kamara, A.Y., Omoigu, L.O., and Karikari, B., 2022. Assessment of the genetic structure and diversity of soybean (*Glycine max* L.) germplasm using diversity array technology and single nucleotide polymorphism markers. *Plants*, 11: 68.

Vandamme, E., Renkens, M., Pypers, P., Smolders, E., Vanlauwe, B., and Merckx, R., 2013. Root hairs explain P uptake efficiency of soybean genotypes grown in a P-deficient Ferralsol. *Plant Soil*, 369: 269–282.

Vu, H.T.T., Kilian, A., James, A.T., Bielig, L.M., and Lawn, R.J. (2015). Use of DArT molecular markers for QTL analysis of drought-stress responses in soybean. II. Marker identification and QTL analyses. *Crop and Pasture Science*, 66: 817–830.

https://doi.org/10.1142/9789811276736_0008

Chapter 8

Digital Transformation on Agricultural Products and Food Markets

Francisco Cisternas

CUHK Business School, The Chinese University of Hong Kong
Institute of Environment, Energy and Sustainability,
The Chinese University of Hong Kong
fcisternas@cuhk.edu.hk

1. Introduction

Having enough food is a balance between demand and supply. Food supply has been a main concern in human society throughout our history. Although food was a general topic in ancient time, it was not until 1798 that the scholar and influential economist Tomas Malthus formalised the observation that while population growth is exponential, our ability to produce food increases in a more linear fashion. This fact will inevitably lead to conflicts, famine, or war, causing poverty and depopulation. These consequences were later called *Malthusian Catastrophe* and followed by many economists and politicians becoming the Malthusianism movement and later the Neo-Malthusian movement that lasts until this day.

Of course, the above proposition has many caveats. Most notably that demand for food cannot increase exponentially forever, or that our ability to produce food has greatly surpassed our demand, thus far. The issue of food shortage persists due to factors such as uneven distribution, limited land for crop cultivation, poverty and more. But digital technologies have emerged as a promising trend in addressing these challenges. Development of new techniques, technology and farming practices in ancient times was knowledge that was hard to preserve, and almost impossible to share. However, this has changed, and now we are able to share and build upon. Information and communication technologies are more powerful than ever in our history, and the results are evident. The creation of digital markets, data-based agriculture, together with improvements in agro-technology is allowing us to implement a sustainable food production taking care of producers, consumers, and the environment. The future of digital transformation in agriculture is potential. By embracing digital tools and technologies, agriculture markets can improve productivity, profitability and sustainability, ultimately contributing to global food security.

In this chapter, we will focus on how we have been able to escape the *Malthusian Catastrophe* for the last centuries. We will center our discussion on the impact of information and communication technologies on the agricultural markets, with an emphasis on developing countries, which produce almost 80 percent of food worldwide.

2. The Challenge

Humankind has existed for about two hundred thousand years and for most of this time, humans were a handful of tribes of hunter-gatherers with populations of a few hundreds. It was only at the end of the last glaciation, around ten thousand years ago, that we were able to sustain larger populations thanks to better weather conditions and the invention of agriculture. Having food around made people stay in the same place, to start living together and form the first villages and cities. From then population growth was stable, except for large wars and occasional large epidemic, also consequences of agriculture, humans thrive expanding territories and outpacing other mammals in terms of usage of land and other resources. In the following five thousand years, population growth only accelerated and big civilisations appeared in what is now East Asia, Europe, and India.

By 1800, the world population reached about 1 billion people; then the Industrial revolution came to speed up everything, and 120 years later the population doubled to 2 billion. It took over hundred thousand years to reach one billion people and only 120 the second billion. Today, 100 years later, we quadruple the population to almost 8 billion (Our World in Data, Worldometer). These numbers are staggering and hard to comprehend, one of every 12 humans that ever lived in 200 thousand years of history is alive today.

However, population growth is slowing down, and it is expected to reach about 10 billion by 2050 and to reach a peak of 11 billion by 2100 according to the United Nations (Desa, 2019). This means we need to produce food for a couple of billion people more in a few decades. It has been estimated that an increase in agricultural production of 60% (based on tonnes by crop prices) is needed by 2050 to meet the food demand of the projected population (Alexandratos and Bruinsma, 2012).

Additionally, Fukase and Martin (2020) showed that each person today is consuming more food than ever before, and the contribution of this factor into food shortages is more important than population growth.

The size of modern cities in terms of numbers as well as physical scale is unprecedented. In 1800, there was only one city with a million people, London. By 1990, the world's 100 largest cities accommodated 540 million people and 220 million people lived in the 20 largest cities, megacities of over 10 million people, some extending to hundreds of thousands of hectares, land that used to be fertile land now is covered in concrete (Deelstra *et al.*, 2000)

Advances in science and technology have enabled food production to yield more than enough food for everyone alive today. However, there are still people with not enough food due to poverty and lack of access to food markets.

3. What is Currently Working

The benefits of fertilisers and pesticides cannot be overstated. They have been crucial in increasing food production.

As one noticeable recent example, in Spring 2021 Sri Lanka's government banned the use of synthetic fertilisers and pesticides practically overnight, forcing millions of farmers to go organic. Just five months later, the disaster was so evident that the measure had to be reverted. Sri Lanka was self-sufficient in rice production and after six months, it had to spend 450 million in rice imports, and the tea production, the country's biggest export, dropped by 18 percent, costing the government millions in subsidies (Cordell *et al.* 2021).

Food production methods that reduce the impact on the environment and increase productivity per area are becoming increasingly important. Factors such as urbanisation of agricultural land and ever-increasing urban migration have determined the need to enhance the efficient use of land resources (Pandey *et al.*, 2015; Lambin *et al.*, 2013).

Vertical farming systems have been proposed as a solution for this problem (Despommier, 2013). The main concept is to stack growing beds on top of one another increasing crop production into the vertical dimension (Resh, 2012). Different implementations of this concept have been done in several countries like Singapore, Japan, and South Korea (Despommier, 2013). This high-end technology offers unique opportunities to countries like Singapore, which has land scarcity and high levels of domestic food demand. Vertical farming is playing an important role for Singapore's food security strategy, which translates in both public and private investment in R&D (Wood *et al.*, 2020). Sky Greens vertical farming system (www.skygreens.com) is one of the successful efforts in Singapore to ensure food supply.

Examples of vertical farming implementations include mobile vertical planting systems to increase light availability (Mahdavi *et al.*, 2012), and vertical soilless systems (Neocleous *et al.*, 2010). A recent study found that vertical farming offers advantages over conventional horizontal hydroponic systems (Touliatos *et al.*, 2016). The vertical system increased the model crop, *Lactuca sativa* L. cv. "Little Gem", yield per unit area.

Food and farming practices are being propelled by the developments in the Internet of Things, cloud computing, and artificial intelligence among others (Sundmaeker *et al.*, 2016; Per-Anders Langendahl, 2021). These technology innovations are allowing the integration of data into the farm management cycle (Wolfert *et al.*, 2017). Smart farming takes advantage of real-time events information to mitigate impacts on food production. Farming productivity can also benefit from information and communication for decision-making such as for the use of fertilisers and pesticides (Carolan, 2016), and weed control (Fennimore, 2017).

Automation technologies and mechanisation using agricultural robots play an important role in these applications. The use of unmanned aerial vehicles has shown promising results to detect value crops on farms (Lotte, 2017). This type of application allows classifying individual plants and differentiating weeds from crops. Heterogeneous ground and aerial robots have been used for the efficient use of pesticides to improve sustainable agriculture (Gonzalez-de-Santos *et al.*, 2017). Other examples include the use of robots to automatically milk cows (Eastwood, 2022) and harvest crops (Kamata *et al.*, 2018). Additionally, farming is not only limited to rural areas and smart farming is showing potential to help urban farming practices (Per-Anders Langendahl, 2021).

Overall, smart farming offers tools to increase efficiency in agriculture. However, these technologies must consider farmer perspectives, functionality, and potential effects on farmers and the daily farm work (Lunner-Kolstrup *et al.*, 2018; Ingram *et al.*, 2022).

4. Importance of Enhancing Digital Markets for Agriculture

One of the major benefits of digital markets and information technologies is that this enables access to larger markets. In a globalised world this means farmers can produce goods to be sold not just within their community but to other towns, states, or any country in the world.

Agricultural product markets in many developing countries are often poorly integrated (Bardhan 1989, Banerjee and Munshi 2004). As a consequence, producers need to incur in high search costs that tend to lower competition and create an inefficient allocation of goods across markets.

When information is limited or costly, both producers and buyers are unable to engage in optimal exchanges. Excess price dispersion across markets can arise, and goods may not be allocated efficiently. This is a serious problem in agriculture that can lead to under investment in production capacity, significant waste of production resources and agricultural products, discouraging agents to enter the agricultural industry, all resulting in an overall decrease in agricultural outputs. Several studies have shown that the adoption of basic communication and information technologies can reduce price dispersion significantly. For example, Jensen (2007) studied the effects of the adoption of mobile phones in the early 2000's in a fishing city in India. This study shows that both sellers and buyers benefited from the enhanced information sharing increasing sellers' profits and consumers' welfare, reducing the number of unserved consumers. This positive effect was also found in Niger, when mobile phones were introduced between 2001 and 2006, reaching a 16 percent reduction in grain price dispersion (Aker 2010).

In Uganda was found that just broadcasting the commodities' prices on FM radio, helped farmers to bargain for better prices on surplus crop production, and reaching areas with unserved customers (Svensson, Yanagizawa 2009)

Additionally, the lack of transparency in agricultural supply chains are often dominated by exploitative intermediaries with substantial market power (Besley and Burgess 2000). Market transparency improves bargaining power with middlemen, and market participation in remote areas through more efficient coordination. Muto and Yamano (2009) found that expansion of information technologies through cell phones enhanced market participation of farmers in remote areas who produce perishable crops.

When internet kiosks and warehouses were established in central India, soy farmers had access to wholesale price information and an alternative marketing channel in other parts of the state. An immediate increase in soy price was found after the introduction of the kiosks. Moreover, the area under soy cultivation also increased, signaling an improvement in the soy market that benefited consumers with more soy production and farmers with higher prices.

Context specific factors and various marketing and institutional constraints can blunt benefits. These factors include small-scale farms, lack of physical infrastructure and insufficient human capital, very common in undeveloped and developed countries, can significantly reduce the benefit of investments and adoption in improving the digital channels and information technologies. Lio and Liu (2006) studied a panel of 81 countries between 1995 and 2000. They found that the adoption of new information technologies had an impact up to two times greater in countries with more resources.

A study of expansion of information technologies in Rwanda (Futch and McIntosh 2009), found limited impact for local farmers measured by changes in price or production levels after the introduction of a village phone. Muto and Yamano (2015) found a similar issue when looking at multiple crops in Uganda, finding that adoption of mobile phones had a positive effect on prices of bananas but not in maize. One potential reason that this effect is more pronounced in bananas is because they are perishable, therefore the freshness at the time of exchange strongly influences its market value. This does not suggest that the mobile phone network does not have significant impact on cereal grains such as maize. Although in Muto and Yamano (2015) was not significant, both Aker (2008) and Goyal (2008) showed the positive effect in these goods.

The context in food production such as physical infrastructure and electricity in rural areas plays an important role in taking advantage of information technologies. It is not enough to have mobile phone coverage, if people in rural areas cannot charge their phones, or reach out to the new markets. Social barriers like language, education level and ethnic segregation also affect the reach of the food markets. For example, in many African countries, women are the main producers of food crops and men are those of cash crops, produced mainly for sale. When women have less access to education than men do, making the adoption of digital solutions to facilitate market access much harder. Finally, regulation also plays an important role in developing food markets.

In terms of productivity, digital markets and information technologies also play a big role, facilitating the adoption of improved inputs by providing advice and weather forecasts at a lower cost and enhancing investment decisions. Best practices in the production can be shared among farmers and the result of investment decisions can be spread easily, adding to the benefit of shared information. A study in which experts were brought to share information with farmers increased adoption of certain agricultural practices up to 7-fold (Toyama, 2009) or a hotline service provided in India with information about seeds and fertiliser as well as crops choices lead to change in investment decisions (Cole and Fernando, 2012). In a similar way in Chile, a cooperative of farmers used text messages to help small farmers to provide planting advice and weather updates, particularly useful in critical conditions like snowing or harvest (George *et al.*, 2011). When this information is readily available, it can also prevent losses, with early warning systems on climate and pest outbreaks (Ceccato *et al.*, 2014). Additionally, these technologies help with automated monitoring systems. In the future, these systems, based on social media or crowdsourcing, will help farmers by providing real-time information in remote areas with limited access to other information sources. Early steps in this direction are seen in Indonesia, where food prices and food inflation correlate with tweets speaking about food prices (Pulse, UN Global, 2014)

Digital markets and information technologies can enhance management of farms. These tools will increase income and value of assets of farmers. Successful examples of the use of these tools includes Philippines (Labonne and Chase, 2009) and Peru (Beuermann *et al.*, 2012). Beuermann (2012) found household consumption rose by 11 percent and poverty incidence was reduced by 8 percent after the introduction of phone coverage in Peru. Although we cannot attribute only to the effect in enhanced agriculture performance, the increased wealth still translates to a more sustainable food production.

Accurate and on time information enables logistics to be more efficient. Effective logistics is critical for producers, retailers, as well as consumers for collection, aggregation, and delivery. Various field studies show that in many places digital technologies are quietly transforming how rural logistics function (George *et al.*, 2011). The resulting improvement in logistics can be seen through lower transaction costs, improved profits, and less wastage.

5. Discussion

Food security is one of the greatest challenges we are facing as a society. We have been working in many domains to address these issues. Human ingenuity so far has outpaced the challenges, but the task is growing with an ever-growing demand, resources depletion and environmental deterioration. Science progress and new technology have led the way improving food production, but we also need to improve food markets and make them sustainable, attending the needs of food producers, using information to connect all corners of the world with the food networks.

As information technologies spread to agricultural areas, especially in developing countries, several studies have shown significant impact on food production and social welfare for farmers. However, there is also evidence that these innovations often fail to scale-up and achieve wider acceptance. Access to information and the ability to communicate are only two of many constraints farmers face. Others include market fragmentation (even though market consolidation will, over time, enhance growth prospects). The lack of financially sustainable business models to attract private sector investments for providing innovative solutions for small scale agriculture. There is clearly high potential for the internet and related technologies to improve rural economies, but several lessons need to be kept in mind.

First, we explore many ways in which new technology can and has improved the quality and quantity of food production benefiting the farmers and consumers. Technology such as Digital soil maps, remote sensing, and GPS guidance are critical tools for modern farmers. Big data for precision agriculture increases yields and efficiency. However, these technologies require large investments and human capital, which makes them only available for large companies. Improvements on technologies also make them cheaper and easier to use, from which smaller farmers can have access to. We saw how simple mobile phones can already make a big difference.

For this to have a widespread impact in rural areas of developing countries requires the closing of the remaining digital divide. While mobile phones have spread quickly even in low-income countries and among poorer population groups, access is by no means universal.

Second, basic price and market information systems can improve efficiency and welfare. Although the evidence is strong, recent studies show that in some cases this may not be enough. When information requires a fee, the effects seem to be weaker, and even when informed some farmers seem to not act on the information because they cannot reach other markets or supply chains. To take a full advantage of the benefits of new technology is necessary a deeper assessment of the needs of farmers.

The World Bank (2016) report on digital dividends put it simply saying that "It is that usually only some part of a task or service can be automated using technology while the remaining part requires non-routine skills such as discretion or complex problem solving that cannot be done by machines". A better understanding of the interplay of what is automatable and what is not could help better explain why some technologies (such as mobile money) take off, while others seem to underperform expectations. Such insights could also help design better interventions that leverage technology for rural development.

Finally, technology itself is not enough, they need to be accompanied by complementary investments in physical infrastructure and enhancement of human capital (Toyama, 2015). When new technologies appear, it is easy to forget that without the proper supporting infrastructure or educated people to make use and take advantage of these new opportunities, the effects may be disappointing. All this needs to be in line with an overall regulatory environment. All of this highlight the key role of the government in providing adequate infrastructure and enhancing the human capital to be increase the speed of adoption of new technologies, and adapt and potentially generate new technology to handle specific circumstances.

Successful implementation of new technologies to enhanced agricultural markets can have a significant role in securing food production and to make it sustainable, for both the environment and producers.

Acknowledgement

The author gratefully acknowledges the insights and comments during the presentation at the Agro-biotechnology talk series. He also likes to thank Carolina A. Contador for her contribution sourcing references and suggestions. Finally, he wants to thank Professor Hon Ming Lam for his motivation, and encouragement, which made this project possible.

References

Aker, Jenny C., 2008. Does digital divide or provide? The impact of cell phones on grain markets in Niger. *Center for Global Development Working Paper* 154.

Bakker, N., Dubbeling, M., Gündel, S., Sabel-Koshella, U., and de Zeeuw, H., 2000. Growing cities, growing food. Urban agriculture on the policy agenda. Feldafing, Germany: Zentralstelle für Ernährung und Landwirtschaft (ZEL). In: Tjeerd, D. and Girardet, H. (eds.) *Urban Agriculture and Sustainable Cities*: pp. 43–66.

Dana, C., Dominish, E., Esham, M., Jacobs, B. and Nanda, M., 2021. Adapting food systems to the twin challenges of phosphorus and climate vulnerability: the case of Sri Lanka. *Food Security* 13(2): 477–492.

Desa, U.N., 2019. World population prospects 2019: Highlights. *New York (US): United Nations Department for Economic and Social Affairs*, 11(1): 125.

Dickson, D., 2013. Farming up the city: the rise of urban vertical farms. *Trends Biotechnol.*, 31(7): 388–389.

Diether, W. B., McKelvey, C., and Vakis, R., (2012). Mobile phones and economic development in rural Peru. *J. Dev. Stud.*, 48(11): 1617–1628.

Douglas, F. M. and Craig Thomas McIntosh, C. T., 2009. Tracking the introduction of the village phone product in Rwanda. *Inf. Technol. Int. Dev.*, 5(3): p. 54.

Eastwood, C.R., Dela Rue, B., Edwards, J.P., and Jago, J., 2022. Responsible robotics design — A systems approach to developing design guides for robotics in pasture-grazed dairy farming. *Front. Robot. AI*: 9.

Emiko, F. and Martin, W., 2020. Economic growth, convergence, and world food demand and supply. *World Dev.,* 132: 104954.

Fennimore, S.A., 2017. Automated weed control: new technology to solve an old problem in vegetable crops. In: *Conference presentation at ASA Section: Agronomic Production Systems.*

Friess, Peter. F., 2016. *Digitising the Industry-Internet of Things Connecting the Physical, Digital and Virtual Worlds.* River Publishers.

George, T. *et al.*, 2011. ICT in Agriculture. Connecting Smallholders to Knowledge, Networks, and Institutions.

Gonzalez-de-Santos, P. *et al.*, 2017. Fleets of robots for environmentally-safe pest control in agriculture. *Precis. Agric.*, 18(4): 574–614.

Goyal, A., 2010. Information, direct access to farmers, and rural market performance in central India. *Am. Econ. Appl. Econ.*, 2(3): 22–45.

Gupta, A., Ponticelli, A., and Tesei, A., 2020. *Information, Technology Adoption and Productivity: The Role of Mobile Phones in Agriculture.* No. w27192. National Bureau of Economic Research.

Ingram, J. *et al.*, 2022. What are the priority research questions for digital agriculture? *Land Use Policy*, 114: 105962.

Jensen, R., 2007. The digital provide: Information (technology), market performance, and welfare in the South Indian fisheries sector. *Q. J. Econ.*, 122(3): 879–924.

Kamata, T., Ali, R., and Noboru, N., 2018. Heavy-weight crop harvesting robot-controlling algorithm. *IFAC-PapersOnLine*, 51(17): 244–249.

Labonne, J. and Chase, R.S., 2009. The power of information: the impact of mobile phones on farmers' welfare in the Philippines. *World Bank Policy Research Working Paper* 4996.

Lambin, E.F. and Patrick, M., 2011. Global land use change, economic globalization, and the looming land scarcity. *Proc. Natl. Acad. Sci.*, 108(9): 3465–3472.

Langendahl, P.-A., 2021. The Politics of Smart Farming Expectations in Urban Environments.

Lio, Monchi, and Liu, M.-C., 2006. ICT and agricultural productivity: evidence from cross-country data. *Agric.l Econ.* 34(3): 221–228.

Lottes, P., Raghav, K., Johannes, P., Roland, S., and Cyrill, S., 2017. UAV-based crop and weed classification for smart farming. *IEEE Int. Conf. Robot. Autom.*: 3024–3031.

Lunner-Kolstrup, C., Torsten, H., and Janne, P.K., 2018. Farm operators' experiences of advanced technology and automation in Swedish agriculture: a pilot study. *J. Agromedicine*, 23(3): 215–226.

Mahdavi, S., Mohsen, K., Rouhangiz, N., and Toktam, S., 2012. Vertical mobile planting system consistent with the pattern of solar radiation and effects of system on light exposure and growth of Gerbera cut flowers (Gerbera jamesonii cv. Antibes), in greenhouse culture. *J. Agric. Technol.*, 8: 1461–1468.

Michael, C., 2017. Publicising food: big data, precision agriculture, and co-experimental techniques of addition. *Sociologia Ruralis,* 57(2): 135–154.

Mitra, S., Dilip, Mookherjee, M.T., and Sujata, V., 2018. Asymmetric information and middleman margins: An experiment with Indian potato farmers. *Rev. Econ. Stat.*, 100(1): 1–13.

Munshi, K., 2004. Social learning in a heterogeneous population: technology diffusion in the Indian Green Revolution. *J. Dev. Econ.*, 73(1): 185–213.

Muto, M. and Takashi, Y., 2009. The impact of mobile phone coverage expansion on market participation: Panel data evidence from Uganda. *World Dev.,* 37(12): 1887–1896.

Neocleous, D., Charalambos, K., Nicos, S., and Polycarpos, P., 2010. Horizontal and vertical soilless growing systems under Cyprus conditions. *J. Appl. Hortic.*, 12: 140–144.

Nikos, A. and Bruinsma, J., 2012. World agriculture towards 2030/2050: the 2012 revision.

Pandey, B. and Karen C.S., 2015. Urbanization and agricultural land loss in India: Comparing satellite estimates with census data. *J. Environ. Manage.*, 148: 53–66.

Pranab, B. (ed.), 1989. *The Economic Theory of Agrarian Institutions*. Clarendon Press, 1989.

Pulse, UN Global, 2014. Mining Indonesian tweets to understand food price crises. *Jakarta: UN Global Pulse*.

Ramankutty, N., Jonathan A.F., and Nicholas J.O., 2002. People on the land: Changes in global population and croplands during the 20th century. *AMBIO*, 31(3): 251–257.

Resh, H.M., 2022. *Hydroponic Food Production: A Definitive Guidebook for the Advanced Home Gardener and the Commercial Hydroponic Grower*. CRC Press.

Runck, B.C., Joglekar, A., Silverstein, K.A.T., Chan-Kang, C., Pardey, P.G., and Wilgenbusch, J.C., 2021. Digital agriculture platforms: Driving data-enabled agricultural innovation in a world fraught with privacy and security concerns. *Agron. J.*

Shawn, C. and Fernando, A.N, 2012. The value of advice: Evidence from mobile phone-based agricultural extension. *Harvard Business School Working Paper* No. 13-047.

Sundmaeker, H., Verdouw, C.N., Wolfert, J., and Perez Freire, L., 2016. Internet of food and farm 2020. In *Digitising the Industry*, Vol. 49: pp. 129–150. River Publishers.

Svensson, J. and Yanagizawa, D., 2009. Getting prices right: the impact of the market information service in Uganda. *J. Eur. Econ. Assoc.*, 7(23): 435–445.

Timothy, B. and Burgess., R., 2000. Land reform, poverty reduction, and growth: Evidence from India. *Q. J. Econ.,* 115(2): 389–430.

Touliatos, D., Dodd, I.C., and McAinsh, M., 2016. Vertical farming increases lettuce yield per unit area compared to conventional horizontal hydroponics. *Food Energy Secur.*, 5(3): 184–191.

Toyama, K., 2015. *Geek Heresy: Rescuing Social Change from the Cult of Technology*. PublicAffairs.

Toyama, K., Gandhi, R., Veeraraghavan, R., and Ramprasad, V., 2009. Digital green: Participatory video and mediated instruction for agricultural. *Inf. Technol. Int. Dev.*, 5(1): 1–15.

Uwe, D., Goyal, A., and Mishra, D., 2016. Will digital technologies transform agriculture in developing countries? *Agric. Econ.,* 47(S1): 21–33.

Wolfert, S., Lan Ge, C.V., and Marc-Jeroen B., 2017. Big data in smart farming — a review. *Agric. Syst.,* 153: 69–80.

Wong, C., Jacob W., and Swathi, P., 2020. Vertical farming: An assessment of Singapore City. *eTropic: Electronic Journal of Studies in the Tropics* 19: 228248.

World Bank Group, 2016. *World Development Report 2016: Digital Dividends*. World Bank Publications.

Wu, X.D., Guo, J.L., Han, M.Y., and Chen, G.Q., 2018. An overview of arable land use for the world economy: From source to sink via the global supply chain. *Land Use Policy* 76: 201–214.

https://doi.org/10.1142/9789811276736_0009

Chapter 9

Intangible Assets of Agrobiotechnology

Albert Wai-Kit Chan* and Ivan Tomislav Crnosija†

Law Offices of Albert Wai-Kit Chan, World Plaza, Suite 408
141-07 20th Avenue, Whitestone, NY 11357, USA
**akitchan@aol.com*
†crnosijai@kitchanlaw.com

1. Introduction

Intangible Assets (IAs) comprise intellectual property, databases, regulatory approval, key opinions, certificates, and many other valuable resources. This chapter aims to describe the IAs of agrobiotechnology. The roadmap is recited below:

- Background of Plant Variety Protection (PVP)
- United States
  - Plant Variety Rights (PVR)
  - Plant Patent
  - Utility Patent
- Other Intangible Assets
  - Regulations
  - Databases
  - Information Networks

2. Intellectual Property Protection of Plants in the United States

The United States offers an assortment of ways to protect plants.

2.1. *Trade Secrets*

Trade Secrets (TSs) are very important for practicing agriculture. There are two important elements for a TS (US Patent and Trademark Office, 2022) — (1) It must be a secret which, if disclosed, is no longer protectable; and (2) The TSs must provide a competitive advantage for the entity practicing it. Entities need to have measures in place to protect their TSs. These include well-guarded procedures (i.e., secret recipes, lock-and-key protection, etc.) which allow companies to continuously practice their TS without disclosing specifics of how it is done. For example, seed breeders and entities that do culturing in their daily practice depend on safeguards to operate their businesses without others copying them.

2.2. *Plant Variety Protection Act (PVPA)*

The U.S. Plant Variety Protection Act (PVPA) provides one avenue of protection. Signed into law in 1970, the PVPA was meant to incentivise investment in the development of new plant varieties (Plant Variety Protection Act, 7 U.S.C. § 2581, 1970). Before the law's enactment, advancements in agriculture depended primarily on public funding. Private funding, meanwhile, stagnated as those who developed novel plant varieties lacked the means to protect their investments. By providing patent-like rights in plant varieties, the PVPA encouraged widespread innovation in the agricultural space (US Congress, 1986).

To be eligible for protection under the PVPA, a plant variety must be sexually reproduced or tuber propagated. Further, it must be new, distinct, uniform, and stable. A variety is new if it has been sold for no more than one year in the United States, or no more than four years outside of the United States. A variety is distinct if it can be distinguished from varieties known to the public based on several characteristics, including but not limited to morphological and physiological characteristics. A variety is uniform if variations are describable, predictable, and commercially acceptable. Finally, a variety is stable if reproduction does not result in changes with respect to its essential and distinctive characteristics, with a degree of reliability comparable to similar varieties (Plant Variety Protection Act, 7 U.S.C. § 2402, 2020).

The PVPA is a voluntary program, meaning that rights in a plant variety do not vest unless someone who breeds, develops, or owns a plant variety files an application with the Plant Variety Protection Office (PVPO) and the PVPO issues a certificate of protection. Such applications involve depositing a sample of propagation material with the PVPO and paying certain fees. If the PVPO issues a protection certificate, the certificate holder may exclude others from selling, offering for sale, marketing, conditioning, stocking, reproducing, importing, and exporting the protected variety, as well as using the protected variety to produce a hybrid or different plant variety (Plant Variety Protection Act, 7 U.S.C. § 2483, 1996). The term of these rights is 20 years from the date of issuance of the certificate, or 25 years for trees and vines (Plant Variety Protection Act, 7 U.S.C. § 2548, 1991). However, the PVPA provides certain exemptions for farmers and researchers. Namely, if a farmer plants a protected variety, he/she may save enough of its seed upon harvesting to replant an area of land equal to the area for which the protected variety seed was originally purchased (Plant Variety Protection Act, 7 U.S.C. § 2483, 1996, 7 U.S.C. § 2543, 2012). Researchers, meanwhile, may use protected varieties or parts thereof to develop new varieties, as distinguished from producing new varieties (Plant Variety Protection Act, 7 U.S.C. § 2544, 1970).

Certificate holders may bring civil action against those who infringe their PVPA-derived rights. If the plaintiff prevails, the court may grant injunctive relief and damages to compensate for the infringement. These damages cannot be less than a reasonable royalty rate for the protected variety, with interest and costs determined by the court. (Plant Variety Protection Act, 7 U.S.C. § 2564, 2011). Further, the court may award triple damages where the defendant's conduct is found to be willful.

2.3. *Plant Patent Act (PPA)*

Protection may also be sought under the Plant Patent Act (PPA). Unlike the PPVA, the PPA only provides intellectual property rights in asexually reproduced plants, excluding tubers. (35 U.S.C. § 161, 2012).

Modes of asexual reproduction include, but are not limited to, cuttings, grafting, and bulbs. Accordingly, the PPA does not afford protection to seeds or plants grown from seeds. To be eligible under the PPA, a plant must also be invented or discovered in a cultivated state, new, distinct (35 U.S.C. § 161, 2012), and non-obvious (35 U.S.C. § 103, 2011), and applicants for plant patents must provide a description of the plant and a variety denomination (37 C.F.R. § 1.163(a), 2012). The owner of a plant patent may exclude others from using the protected plant, offering it for sale, selling it, importing it, or reproducing it asexually for a period of 20 years from the date their application was filed (35 U.S.C. § 154, 2015) (35 U.S.C. § 163, 1952).

2.4. *Utility Patents*

Plants may be protected like non-plant inventions by acquiring a utility patent. Only "new and useful" plants are patentable (35 U.S.C. § 101, 1999).

Plants that are not novel or those that can be anticipated by others are unpatentable. Plants that are non-operative or lack any practical use are similarly patent ineligible.

> Major goals of patenting include acknowledging inventors' discoveries and innovations and promoting the progress of technological development (35 U.S.C. § 101, 1999). Therefore, only new plants would be considered patentable inventions. Even so, a new use or improvement of a known plant may be patentable if it entails an innovation and bestows beneficial outcomes upon the community.

USPTO examiners determine the novelty of an invention by determining whether the claimed invention has been described in prior art, which is all information available to the public related to the invention (35 U.S.C. § 102, 1999). The novelty requirement is rarely a tough hurdle to patentability, as only prior art that is identical to the claimed invention would destroy the novelty of the invention. Conversely, objections over obviousness, which is equivalent to a lack of inventiveness, are usually expected and not so easily overcome.

Examiners judge whether an invention is inventive by considering whether prior art provides any suggestion or teaching such that an ordinary person in the art can achieve or anticipate the same invention by combining the elements in the prior art. Unlike the doctrine of novelty, a combination of multiple prior art can render an invention obvious and therefore unpatentable (35 U.S.C. § 103, 1999). An invention must be significantly different from the prior art and inventors must demonstrate that the invention has produced unexpected results in view of prior art.

Furthermore, an invention would be considered obvious even if it was merely obvious for one to try combining various "identified, predictable solutions with a reasonable expectation of success" to pursue an invention. *In re Wright*, 999 F.2d 1557, 1561 (Fed. Cir. 1993).

> § 112 of the U.S. Patent Law requires applicants to give a fully enabling written description of the invention (35 U.S.C. § 112, 1999). It is acceptable to provide prophetic examples if the ordinary person in the art can achieve the same invention by referring to the disclosure in the application. This requirement is especially difficult for plants, and often requires the deposition of a biological sample with the USPTO (Knauss et al., 2018).

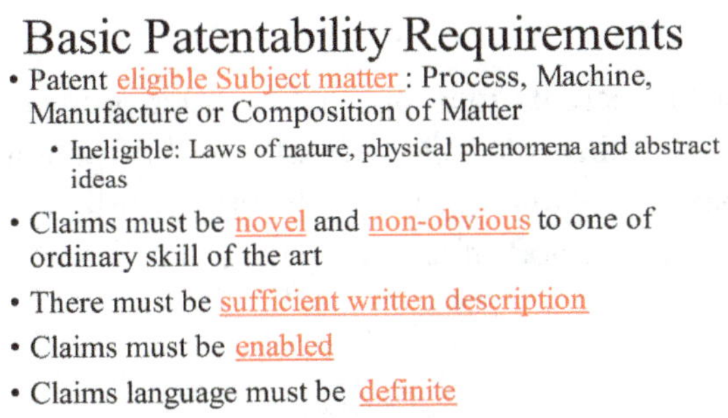

Figure 1. Criteria for the grant of a patent, i.e. patentability.

Utility Patent

- Patentability requirements: utility, new, non-obviousness, written description, enablement, best mode
- Possible to protect a class of varieties with specific traits, plant parts, and methods of producing or using plants
- 20 years from date of filing
- Right to exclude others from making, using, selling, offering for sale and importing the claimed invention in the U.S.

Figure 2. Features of a United States utility patent.

Utility Patent

- Possible to protect:
  - Modified plant genes, protein, products
  - Transgenic plants
  - Class of varieties with specific traits
  - Plant variety
  - Plant parts-cells, tissues, etc.
  - Methods of producing or using plants/varieties etc.

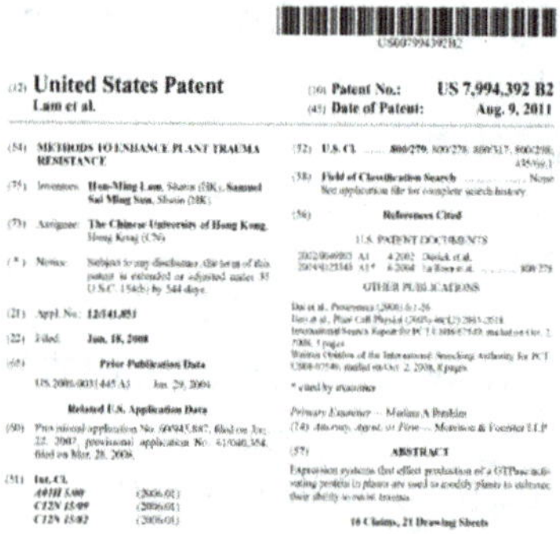

Figure 3. Plant-related inventions eligible for utility patent protection.

Unlike the PPVA or plant patents, utility patents can protect plants regardless of their method of reproduction, i.e. sexual vs. asexual reproduction. Utility patents can also provide protection for plant parts, such as fruits and flowers, which are excluded under the PPA (35 U.S.C. § 161, 2012).

As such, utility patents may provide broader protection for a given plant compared to other IAs. Owners of a utility patent may exclude others from making, using, selling, offering to sell, or importing the protected plant for a period of 20 years from the filing date (35 U.S.C. § 154, 2015).

3. Strategies: Plant Patent vs. Utility Patent vs. Plant Variety Protection

As discussed above, Plant Patents, Utility Patents and Plant Variety Protection (PVP) protect different aspects of plant-related intellectual property. However, these three types of IP rights are not mutually exclusive. Accordingly, if a plant is eligible for protection under more than one of these IP rights, owners should pursue protection strategically (J.E.M. AG Supply, Inc. v. Pioneer Hi-Bred Int'l, Inc, 2001).

3.1. *Plant Patent vs. PVP*

Seed-propagated hemp is an interesting example of a plant invention because federal prohibition has prevented the cannabis industry from obtaining IP rights pertaining to this subject matter until recently. The U.S. Department of Agriculture announced on April 24, 2019 that the Plant Variety Protection Office (PVPO) would begin accepting applications for seed-propagated hemp for plant variety protection (US Department of Agriculture Agricultural Marketing Service, 2019). At the time of this writing, 15 applications have been granted (https://apps.ams.usda.gov/CMS/CropSearch.aspx, 2022). At the same time, IP protection has been granted for hemp since 2016 through the Plant Patent Act of 1930 (Kubby, S.K., U.S. Plant Patent No. 27,475). It is interesting to note that the cannabis industry may find the plant patent to be more valuable than plant variety rights. As discussed previously, the owner of a plant patent may exclude others from using the protected plant, offering it for sale, selling it, importing it, or reproducing it asexually for a period of 20 years from the date the plant patent application was filed. It makes no difference whether the infringement action is for private or commercial purposes. However, in the case of PVP, the act of using the protected plant material or seeds for private, non-commercial use is not considered infringement. Accordingly, PVP rights would not provide sufficient protection for a plant invention that end users use widely for self-consumption.

Strategies – Plant Patent or PVP?

Plant Patent

✓ Higher degree of Control – via homogeneity/stability requirement

X Higher Costs

X Limited to Specific Variety described in Plant

Plant Variety Protection (PVP)

✓ Broader scope of protection – includes Essentially Derived Varieties

Figure 4. Advantages and disadvantages of Plant Patent protection compared to that of PVP.

3.2. *Plant Patent vs. Utility Patent*

While plant patents and utility patents are both "patents," subtle differences distinguish them from one another. Notably, the concept of "invention" varies significantly between the two. A basic requirement for patentable subject matter for a utility patent is that the invention must be "man-made" (Diamond v. Chakrabarty, 1980). However, this requirement would severely limit the plants that would qualify as inventions if the same criteria existed for plant patents. For plant patents, the act of identifying a mutation, isolating it, and reproducing a new variety or sport that is bred via natural means is sufficient to qualify the plant as patentable subject matter (US Patent and Trademark Office, 2017). Further, the requirement to fulfill written description and enablement under 35 U.S.C. § 112 for utility patents is not applicable to plant patents and, under the Plant Patent Act 35 U.S.C. § 161–164, the corresponding requirement is much more relaxed. The description for a plant patent should be as complete as reasonably possible and must include at least one photograph of the new variety and a detailed botanical description of the plant, reciting the genus, species, varietal name, and method of asexual reproduction of the new variety. However, it does not require a person of ordinary skills in the art to make and use the invention or to particularly point out and distinctly claim the subject matter which the inventor regards as his invention. Newly developed plant species therefore fulfill such requirements for a plant patent more readily than those requirements for a utility patent.

Strategies – Plant or Utility Patent?

Plant patents

✓ Less costly and complicated

X One Claim

Utility patent

✓ Stronger Protection – Infringed if asexually/sexually reproduced

✓ Multiple Claims

X Difficult/Time-consuming to obtain

Figure 5. Advantages and disadvantages of Plant Patent protection compared to that of Utility Patent Protection.

4. Interplay of the Three Systems

The best form of IP protection for a plant invention will, first and foremost, depend on the plant species. If eligible, a plant invention often requires all three forms of IP rights for full protection.

Utility patents allow broad protection as defined by the set of granted claims which provide rights to exclude others from making, using, offering for sale, selling, or importing the patented invention. Since the protection conferred depends on how the claims are drafted, the utility patent provides the flexibility to block others from exploiting limitations in PVP or plant patents. For example, one of the main drawbacks of PVP is the farmer's exception which allows the saving of seeds for replanting. Right holders may block this loophole by acquiring utility patent protection for a method of growing or propagating the plant invention. The combination of the utility patent and PVP would not leave a chance for replanting, and the invention would be more comprehensively protected.

Although utility patents provide flexibility in the scope of protection, a biological deposit may be required to fulfill the written description or enablement requirements, which may be a problem for the would-be patent owner. Once a utility patent has been issued, the biological deposit will be made available to the public. Notably, while a utility patent provides only territorial rights, the related biological deposits are available to all competitors including those in unprotected jurisdictions. To prevent competitors from tapping the fruits of labour directly, while still fulfilling the patentability requirements, a utility patent claiming the desired inventive feature may be based upon embodiments of lesser commercial value and its biological deposits submitted. The embodiments having higher commercial value can be protected with a plant patent, which does not require a deposit, or PVP, which only makes deposits available after expiration. In this manner, even though there would be jurisdictions where the plant invention is unprotected, competitors would only have access to the embodiment of lesser commercial value via utility patent biological deposits, and the embodiments of higher commercial value would remain secret.

From an enforcement perspective, having multiple forms of IP rights covering a plant invention provides several advantages. Among the three forms of IP rights discussed herein, the utility patent is the most costly and time-consuming to obtain. Patent infringement disputes are often accompanied by patent invalidation proceedings wherein the defendant tries to counter the infringement allegation by calling the validity of the utility patent into question. Accordingly, there is a risk of losing a utility patent when it is enforced. This risk can be avoided when the patent owner owns PVP rights, and only wields the utility patent when those PVP rights fail to protect the plant invention. In combination with the strategy in the previous paragraph, a utility patent covering a central inventive feature can be bundled with multiple PVPs for each embodiment of high commercial value to form an IP portfolio potential infringers.

5. Regulations

The focus of a company is important. If a company desires to enter the consumer market, it must not only consider the challenges surrounding research and development, but address quality control concerns and navigate consumer-level regulations. By familiarising themselves with geographic regulations early on, companies may avoid costly and time-consuming regulatory hurdles, and potentially increase the value of their products.

In the United States, downstream agricultural products are primarily regulated by two agencies, the United States Department of Agriculture

(USDA) and the Food and Drug Administration (FDA). Approval from these agencies is itself a valuable IA. The USDA enforces standards for quality, also known as "grades," of agricultural products, whereas the FDA enforces standards for product identity. Take, for example, products derived from oranges. The USDA provides that oranges must meet certain requirements with respect to colour, firmness, texture, maturity, similarity of varietal characteristics, be well formed, and that they be free from various imperfections to be classified as "U.S. No. 1." Grade (21 C.F.R. § 146.135, 2011). Under FDA regulations, meanwhile, orange juice must come from mature oranges of *Citrus Sinensis* family or Ambersweet oranges, a citrus hybrid (21 C.F.R. § 146.135, 2011).

The European Union regulatory landscape for downstream agricultural products is comparatively more complex. Many may be familiar with the stringent regulation of wine production and marketing in Europe. Such stringency is pervasive, however. Continuing with orange juice as an example, under the EU Fruit Juice Directive 2001/112/EC, only juice extracted from *Citrus Sinensis* may be sold as orange juice in the EU (Council of the European Union, 2001).

6. International Standards

Similar regulation exists at the international level. The Codex Alimentarius, established by the Food and Agriculture Organization (FAO) of the United Nations (UN) and the World Health Organization (WHO), establishes certain food standards and provides related guidelines and codes of practice. Currently, there are 189 members of the Codex Alimentarius Commission (CAC), which oversees the Codex Alimentarius. Through the Codex Alimentarius, the CAC promotes food safety in member states and encourages food trade amongst them. Member states may only adopt stricter codes than those set out by the Codex Alimentarius with adequate scientific justification. As such, the Codex Alimentarius provides baseline food standards, but discourages unnecessarily strict regulations which may hamper or disadvantage certain member states on the international food trade stage (Food and Agricultural Organization of the United Nations, 2022).

7. Databases and Information Networks

The value of data is indisputable in the modern era. Agrobiotechnology data is no exception. Before gene sequencing technology matured, biological specimens of plant varieties were of secondary interest to players in the agrobiotechnology space. These players were more interested in the advantages and disadvantages of the plant variety itself. Today, genetic information is of the utmost interest. A gene sequence can illuminate the mechanism responsible for a plant variety's value, and greatly inform further developments. Various nations and international organizations maintain public databases of plant varieties. Agrobiotechnology actors would do well to collect and digest as much of this data as possible, and assemble their own private databases, whether they do so in furtherance of developing new plant varieties or for the database's intrinsic value.

Information networks provide a high-level view of advances in the agrobiotechnology space. Unlike the aforementioned databases, which generally focus on the minutiae of individual plant samples, information networks have taken it upon themselves to disseminate modern applications of agrobiotechnology. AgBiotechNet, for example, aggregates scientific literature on the subject and promotes developments to researchers and policymakers (Centre for Agriculture and Bioscience International, 2021).

8. Plant Variety Protection Outside of the US

While patents are inherently limited in their geographic scope, international options exist for expanding the protection of agribiotechnology. The International Union for Protection of New Varieties of Plants (UPOV) manages one of the most important treaties in this space. The UPOV provides standards adopted by over 70 counties, including the US, the EU and China (International Convention for the Protection of New Varieties of Plants (UPOV), 2024). In order to be eligible for registration with UPOV, a plant variety must be novel, distinct, uniform, and stable (International Convention for the Protection of New Varieties of Plants UPOV, December 2, 1961).

In the EU, varieties of all botanical genera and species, including hybrids, may be registered with the Community Plant Variety Office, and owners may exclude others from selling, marketing, conditioning, stocking, offering for sale, offering for reproduction, importing, exporting, or using the variety to produce a hybrid or different variety (Council of the European Union, 1994).

In China, like in the US, plant varieties must be novel, distinct, uniform, stable, and properly named to receive protection. However, only genera and species in China's National List of Protected Plants are eligible for protection. As such, protection is available for a comparatively narrow swathe of plant varieties. The managing authorities in China are the Ministry of Agriculture and Rural Affairs (MARA) and the State Forestry and Grassland Administration (SFGA). Owners may exclude others from producing, propagating, selling propagating material, or using the propagating material of a protected variety in the repeated production of the propagation material of another variety. As with the PVPA in the US, China recognises exceptions for scientific research and farmers producing seeds for their own use (State Council of the People's Republic of China, 1997).

Latin America accounts for 16% of the world's food and agricultural exports and represents an important arena for intangible agrobiotechnology assets (Duff, 2015). Among all producers of genetically engineered (GE) crops in the world, Brazil is the second largest. The country's major agricultural products include soybeans, cotton, coffee, and tobacco, of which the US and China are both significant importers. Brazil exported a total of US$31 billion in agricultural products to China in 2018 (Silva, 2019).

Following the United States and Brazil, Argentina is the third largest producer of GE crops in the world. Argentina farms roughly 12 percent of the world's total GE crop (Yankelevich, 2021). Given the nation's reliance on the Chinese export market, Argentina collaborates closely with China regarding GE crops. For instance, in Argentina, soybean crops cannot be commercialised unless they are variants approved by the Chinese government (Yankelevich, 2021).

Like Argentina and Brazil, Colombia, Mexico, and Peru are signatories to UPOV, and maintain at least the treaty's basic requirements for plant variety protection, i.e., that a plant variety must be novel, distinct, uniform, and stable to be protectable (UPOV, 2024). However, infringing acts and repercussions for plant variety infringement vary amongst these countries.

As one might expect under UPOV, registrants in Argentina, Brazil, Colombia, Mexico, and Peru, may exclude others from selling, producing, and reproducing their plant variety (UPOV, 1978). Thus, they share a baseline definition of infringing activities. However, of these nations, Peru alone has acceded to the terms of UPOV as revised in 1991, which generally grant more rights to plant breeders, and confer stronger protection for plant varieties (UPOV, 2024).

All five countries maintain farmers' and breeders' exceptions. (Baker Mckenzie, 2021). Similar to farmers under the PVP regime in the United States, farmers in these countries may save a certain amount of protected seeds for replanting on their own land, with certain restrictions. Breeders, meanwhile, may use protected varieties to develop new varieties, subject to limitations. For instance, under the comparatively strict regime of Peru, breeders require authorisation from a plant variety owner where the new variety "conforms to the initial variety in the expression of the essential characteristics that result from the genotype or combination of genotypes of the initial variety", even if the new variety is "is clearly distinguishable from the initial variety" (UPOV, 1991).

As infringing acts vary between Argentina, Brazil, Colombia, Mexico, and Peru, so to do potential consequences for infringement. Perhaps most notably, infringers of plant variety protection in Colombia and Peru may face criminal in addition to civil liability. (Colombia, 2006) (Peru, 2003). So, while plant variety protections have been historically underenforced in parts of Latin America, infringers may face significant consequences (The Berne Declaration, 2014).

9. Summary

In the United States the European Union and other developed countries, private sectors dominate most, if not all, agrobiotechnologies. Most developing and underdeveloped nations rely on their governments to serve as sponsor, facilitator and regulator because the private sector in a developing nation may not be able to recoup its investment. Questions remain regarding how to link developed, developing, and underdeveloped nations to create a more efficient ecosystem for facing global issues. The US Agency for International Development (USAID) offers a Global Food Security Grant (GFSG), which is a good start; however, extensive communication with prospective stakeholders and partners should be established. Global programs must be created to define key challenges and offer new solutions. Agrotechnology's intangible assets have been and will continue to play a vital role in all these activities.

Glossary

Claims: Define the scope of a patent, i.e. what it protects.
CAC: Codex Alimentarius Commission
Codex Alimentarius: Sets forth standards, guidelines, and codes of practice with respect to food.
CPO: Community Plant Office
FAO: Food and Agriculture Organization
FDA: Food and Drug Administration
GFSG: Global Food Security Grant
PPA: Plant Patent Act
PVPO: Plant Variety Protection Office
PVP: Plant Variety Protection
PVPA: Plant Variety Protection Act
PVR: Plant Variety Rights
Prior Art: All information available to the public in relation to an invention.
USAID: United States Agency for International Development
USDA: United States Department of Agriculture
USPTO: United States Patent and Trademark Office
TS: Trade Secret
WHO: World Health Organization

Acknowledgements

The authors gratefully acknowledge the insights offered by Dr. Yee Loi Chan and would also like to thank Ms. Charis Lam, Esq. for doing most of the research for this chapter.

References

21 C.F.R. § 146.135 (2011)

35 U.S.C. § 101 (1999)

35 U.S.C. § 102 (1999)

35 U.S.C. § 103 (1999)

35 U.S.C. § 112 (1999)

35 U.S.C. § 161 (2012)

35 U.S.C. § 163 (1952)

37 C.F.R. § 1.163(a) (2012)

85 C.F.R. § 51.682 (2020)

Baker McKenzie. (2021). Plant variety rights summary. Retrieved from https://www.bakermckenzie.com/en/insight/publications/guides/-/media/files/insight/publications/2020/05/baker-mckenzie-plant-variety-rights-summary.pdf

Centre for Agriculture and Bioscience International. (2021, August 15). *Our history.* CABI.org. https://www.cabi.org/about-cabi/our-history/

Colombia. (2006, June 22). Amendment to Article 306 of the Criminal Code by Law No. 1032.

Council of the European Union. (1994, January 9). *Council Regulation (EC) No 2100/94 of 27 July 1994 on Community plant variety rights.*

Council of the European Union. (2001, December 20). *Council Directive 2001/112/EC of 20 December 2001 relating to fruit juices and certain similar products intended for human consumption.*

Diamond v. Chakrabarty, 447 U.S. 303 (1980).

Duff, A. D., and Padilla, A. P. (2015, September 28). Latin America: agricultural perspectives. Rabobank. Retrieved from: https://relooney.com/NS4540/0000-LA-Important_9.pdf

Ex parte Hibberd, 227 USPQ 443 (PTO Bd. Pat. App. & Int. 1985).

Food and Agricultural Organization of the United Nations. (2022, August 29). *About Codex Alimentarius.* https://www.fao.org/fao-who-codexalimentarius/about-codex/en/#c453333

In re Wright, 999 F.2d 1557, 1561 (Fed. Cir. 1993).

International Union for the Protection of New Varieties of Plants (UPOV). (1978). International Convention for the Protection of New Varieties of Plants of December 2, 1961, as revised at Geneva on November 10, 1972, and October 23, 1978. UPOV Publication No. 221(E). Retrieved from https://www.upov.int/edocs/pubdocs/en/upov_pub_295.pdf

International Union for the Protection of New Varieties of Plants (UPOV). (1991). Article 1.4.1 of the UPOV Convention. In *UPOV Convention (1991 Act).* Retrieved from https://www.upov.int/edocs/pubdocs/en/upov_pub_221.pdf

International Union for the Protection of New Varieties of Plants (UPOV). (2024, February 2). *STATUS IN RELATION TO THE INTERNATIONAL UNION FOR THE*

PROTECTION OF NEW VARIETIES OF PLANTS (UPOV). https://www.upov.int/export/sites/upov/members/en/pdf/status.pdf

J.E.M. AG Supply, Inc. v. Pioneer Hi-Bred Int'l, Inc., 534 U.S. 124 (2001).

Knauss, D. J., Veitenheimer, E. E., and Pomeranz, M. (2018). Protecting plant inventions. *Landslide*, 11, 63.

KSR Int'l Co. v. Teleflex Inc., 550 U.S. 398 (2007).

Kubby, S.K., U.S. Plant Patent No. 27,475. Washington, DD: U.S. Patent and Trademark Office.

Peru. (2003, December 13) Law No. 28126, Sanctioning Infringements of Plant Variety Rights.

Plant Variety Protection Act, 7 U.S.C. § 2402 (2020)

Plant Variety Protection Act, 7 U.S.C. § 2541 (2002)

Plant Variety Protection Act, 7 U.S.C. § 2543 (2012)

Plant Variety Protection Act, 7 U.S.C. § 2544 (2012)

Plant Variety Protection Act, 7 U.S.C. § 2548 (1991)

Plant Variety Protection Act, 7 U.S.C. § 2564 (2011)

Plant Variety Protection Act, 7 U.S.C. § 2581 (1970)

Search conducted on August 30, 2022 with "Hemp" as crop name https://apps.ams.usda.gov/CMS/CropSearch.aspx

Silva, J.S. (2019, December 12). Agricultural Biotechnology Annual. United States Department of Agriculture Foreign Agricultural Service. https://apps.fas.usda.gov/newgainapi/api/Report/DownloadReportByFileName?fileName=Agricultural%20Biotechnology%20Annual_Brasilia_Brazil_10-20-2019

State Council of the People's Republic of China. (1997, October 1). *Regulations of the People's Republic of China on the Protection of New Varieties of Plants.*

The Berne Convention. (2014). *Owning Seed, Accessing Food: A Human Rights Impact Assessment of UPOV 1991 Based on Case Studies in Kenya, Peru, and the Philippines.* Public Eye. https://www.publiceye.ch/fileadmin/doc/Saatgut/2014_Public_Eye_Owning_Seed_-_Accessing_Food_Report.pdf

The Berne Declaration. (2014). *Owning Seed, Accessing Food: A Human Rights Impact Assessment of UPOV 1991 Based on Case Studies in Kenya, Peru and the Philippines.* Public Eye. https://www.publiceye.ch/fileadmin/doc/Saatgut/2014_Public_Eye_Owning_Seed_-_Accessing_Food_Report.pdf

United States Patent Department of Agriculture Agricultural Marketing Service. (2019, April 24). *Now Accepting Applications of Seed-Propagated Hemp for Plant Variety Protection*. Agricultural Marketing Services. https://www.ams.usda.gov/content/usda-now-accepting-applications-seed-propagated-hemp-plant-variety-protection

United States Patent and Trademark Office. (2017, September 22). *General Information About 35 U.S.C. 161 Plant Patents*, https://www.uspto.gov/patents/basics/types-patent-applications/general-information-about-35-usc-161

United States Patent and Trademark Office. (2017, September 22). *General Information About 35 U.S.C. 161 Plant Patents,* https://www.uspto.gov/patents/basics/types-patent-applications/general-information-about-35-usc-161

United States Patent and Trademark Office. (2022, July 18). *Trade secrets/regulatory data protection*, https://www.uspto.gov/ip-policy/trade-secret-policy

US Congress, Office of Technology Assessment, Technology, Public Policy, and the Changing Structure of American Agriculture, OTA-F-285 (Washington, DC: U.S. Government Printing Office, 1986).

Yankelevich, A. (2021, January 26). *Agricultural Biotechnology Annual. United States Department of AgricultureForeign Agricultural Service.* Retrieved from https://apps.fas.usda.gov/newgainapi/api/Report/DownloadReportByFileName?fileName=Agricultural%20Biotechnology%20Annual_Buenos%20Aires_Argentina_10-20-2020).

https://doi.org/10.1142/9789811276736_0010

Chapter 10

Farming and I

Miguel Lee Chang

GLCH, Inc.
mlchang@grupoleechang.com

1. The Social Role of Farming

Unlike many other industries, traditional farming is totally outdoors and exposed to all kinds of natural, human, and commercial calamities. Even in the 21st century, farming is still a very primitive task in most part of the world, without being able to generate reasonable income to those who spend their entire life feeding others.

Since thousands of years ago, farming has been family run. A merely 10% of farms is administrated by corporations. With an average farm size of about five acres, (approximately two soccer fields), it is estimated that there are 600 million farms global wise. Without being in the mind of the consumers, farming industry provides over one billion direct job positions, which represent 1/8 of the world population.

The duties of farmers are not limited to producing food for the entire world, but also many other essential necessities of mankind. Other than production of traditional perishable vegetables, crops, meats, and their derivatives, farmers also produce medicines, fiber products for clothes, and structural materials for shelters. Moreover, farming emits oxygen without issuing invoices to anyone, so that the world can keep on rotating cleaner, fresher and smoother.

Farming has always been a motor that avoids urban relocations that potentially turn 'people to be extra social loads for the urban governments. People have seldom noticed another important role of farming, is that being a social stabiliser of rural immigration by holding family integrity. Urban relocation usually starts unchaining families by generating single parent. The vicious poverty swirl keeps the kids from receiving a formal, full, and integral formation due to the paternal absence, without limiting to infrastructural, technologic and academic accesses.

Farming provides not only a main source of financial support for people who live in remote areas, but also is a seed that generates unlimited economic opportunities for many other industries. Even though staying far away from urban areas, farming has always been the foundation of the economic pyramid through land leases, seeds and fertilisers consumption, machineries and infrastructures setup, etc. The chain stretches farther to bring incomes for different sectors including refrigeration equipment and storages, distributions, logistics, wholesale and retail sales, food and beverage, finance, and all the supportive industries derived from farming. More and more academic programs are being offered to innovate the farming industry, food safety and productions related specialties.

There is no discrimination at all in farming. There is no differences in age and gender. Neither has to do with colours, nor believes. The daily duties of farmers usually start before the morning air is heated up by the sunrise, just to assure the produces be evacuated from the farm and sent to the markets before their first clients show up.

2. The Major Challenges of Farming

Among a series of barriers, farmers also have permanent challenges against natural phenomenon. Unstable prices and shrinking margins are almost everlasting placed at the top of the list among multiple adversities.

Escalating regulatory compliances, environmental demands, water availability, shortage of qualified labour, production and financial cost, changing consumer patterns, geopolitical volatility, are among many, many other adversities that farmers have been facing. Heavy bureaucratic structures by centralising governmental policies and technological assistances, risks management, have been keeping many farmers from fair competitions.

While farmers are always subdued to the goodwill and mood of the distributors, supermarket chains can easily establish the final prices according to their operational costs. With scattered locations, farmers are not united enough to defend their own interest for fair and just commercial conditions. The necessity to strengthen their negotiation power, which is traditionally skewed towards the distribution centres, cannot be achieved without being more corporative.

In additional to the hostile commercial environment and disregarding the adverse weather conditions, farm workers need to protect themselves from being attacked by insects, reptiles, and other natural threats, by covering themselves as much as possible, with apron, gloves, cap, scarfs, mask, glasses, boots, etc. That is the daily show that farmers start "enjoying" and "being proud of" since childhood, and for the rest of the life as a rural guardian. Sun burns and snow burns? For sure. They are part of the game. Wrinkles? Well, that is a natural perk.

Farm workers are compromised to work long hours to bring the produces out from the farmland. All family members including kids, women, and seniors, "share the joy" to get the harvest out timely. Extra hands are usually hired to speed up the harvest and avoid losses. There are neither days, nor distances too long. There is no too heavy a load, which is not determined by the physical condition of the farm workers, but their economic necessities.

Farm workers are usually paid not by hour, not by day, but by the production of units. There are no day off during sowing and harvest time. The more one produces, the more one earns, but still, the incomes are far from being enough to support a decent life of a family. In general, farm workers and especially those in developing countries, are not protected by social security, insurances, and other benefits.

For those cases that harvest cannot be mechanised because mostly of the steep terrane, produces are brought out from the farmlands by foot and through long, dangerous, and slippery trails shaped out by the farm workers. Good exercises? Indeed, they are. That is why one hardly can find any overweight farm workers.

In most of the cases global wise, it is an outstanding priority for the governments to improve the basic infrastructures that facilitate farmers to bring their produces out in the shortest possible time.

The limitation of qualified labour force represents another latent burden. Throughout the years with a broader range of job positions and a better furnished lifestyle, cities have been a 2nd choice for some young people. For the farmers, it is a duty to cultivate food that people line up for in the markets, and also to cultivate a new generation to secure the relay for the food production chain.

By blaming farmers for environmental responsibility, consumers are irresponsible. Consumers do not realise that their progressive demand of meats has been cornering the land's capacity to a limit. The chronic unconsciousness from a growing consumer population that demands more food, has pushed the farmers to overgrase by devastating forests that provokes erosions and shortening water supplies. Cattles are no longer raised in the grassland, but confined. Poultries are sent to the markets much earlier than they used to be. Methane emission by live stocks has been accelerating the greenhouse effect that switches the red light on for the humanity.

Having a time like never before, farmers are confronting serious and upgrading challenges, without limited to internal factors, but also international constraints. Global warming has been altering the natural climate patron. Geopolitical situations have slowed down exportations with higher logistic tariffs, on one side; and on the other, elevate the financial cost to operate the farms.

Due to the unpredictable climate complications, farmers have been getting more stressful to comply with additional regulations that pile up the production costs that keep on reducing their shrinking profit. As a result of financial instability, farmers hardly can be qualified to get loans from the banks to make the ends met. For the same reason, insurance coverage is totally absent.

It is astonishingly unbelievable that the banking system, prefers to give out loans to someone unknown for ostentatious vacations, instead of lending the money to those who produce food 24/7 for everyone that includes the bankers!

3. The Production and Waste of Food

Food shortage is not an exclusive fairy tale for developing countries, but also for many first world communities. The main common denominators are governmental bureaucracy and logistic deficiency. While farmers keep on producing food, logistic facilities are not catching up. Technologies and financial support are out of reach. Despite of the expanding world population, food has never been short, but the speed of the logistic chain has not been matching. Consumers keep on draining food without self-containment.

According to different research, 1/3 of the food that come out from farmlands ends up as trash in a way or another. Some produces are simply left in the field without being harvested because the market prices are not enough to cover the expenses. A huge quantity of food is wasted along the logistic chain, starting from the cold chain composed of transportation and storages, exhibition shelves, kitchens, expirations, and worse of all, on the plate of the consumers, both at home and eat out.

When people order their food, fancy or not, they are not aware of the duty to have them finished. In the cases of self-service events, situations turn out even worse, by going for another plate without having tried lightly the previous one. Hundreds, if not thousands, of tons of food are wasted daily in avoidable ways. If consumers can be more self-disciplined, and be mercy enough, hunger could have been contained since long ago on this planet, because food has been shared unevenly by eight billion people.

It is so stunning that an estimated 800 million people in the world suffer from obesity, by eating more than what they need, and by taking away the ration that many others are in daily need. Situation is even more desperate when enormous amount of money is spent on medications and treatments to get rid of the extra pounds that they never have needed. It is not easy at all to comprehend the human behaviour by leaving foods derogatorily on the plate served by the waiters, who are the first ones in line that cannot afford to buy.

Devastating and coincidentally, another 800 millions of people in the same world, are undernourished or suffering from hunger, without being able to access for not even drinkable water. Thousands die every single day. Day after day, since the very beginning of the human civilisation, without able to find a way to be fed. The limitation for these people to access many basic resources has prevented them from escalating in our social ladder, of which all of us are believed to have the same right.

If we are not conscious enough with those in need, we will end up facing so huge a bill ahead that we simply will not be capable to settle.

The ideal is to have the food from the farm to the plate and from the plate to the stomach, instead of from the farm to the trash.

4. The Satisfaction of Being a Farmer

There is a concept that quotes: "If you want to be a public servant, do not expect to have too big an account in the bank. Otherwise, you'll be deeply disappointed". Something similar happens if one is a farmer.

The mission of a farmer is to feed the world by producing food and generating well-being for the society, instead of focusing only on one's financial yield. The duty of a farmer is also to be a rural developer, an environment guardian, an economy driver, and an employment generator. The essence to comply the sacred mission, cannot and should not be measured monetarily.

To be a farmer is a life compromising decision. The joy to provide a better quality of life by feeding others, is an enormous and unique satisfaction. For better, for worse; for richer and for poorer, the duty of a farmer is not to make a fortune, but a living for one's family in a simple, healthy, natural, and harmonious environment. Farmers are also blessed to stick to their family members, with whom one shares daily.

The world is in needs of farmers, and not the other way around. No matter how true this may be, people are celebrating days dedicated to all family members, professions, countries, and all kinds of festival. Without any day declared as Farmer's Day, every day is Farmers' Day, simply because we need to be fed every single day, and not only once, but several times a day.

Differently to the nomads, who are in constant relocations, farmers are the most faithful guardians of the nature, and assure that the world can be accordingly fed. While doctors cure illnesses, farmers prevent diseases. Against all kind of volatilities that may have been diminishing the food industry, farmers have been working harder than ever to avoid it from happening.

5. The Future of Farming

With no doubt, the future of farming will not be in the traditional way, but more demanding and with the participation of more young people, who certainly come up with different mindsets and better technologies to achieve efficiency, precision, and innovations. Food safety is crucial by restoring the continuous damages that this planet has been suffering.

The future of farming is not single dimensional, but differs among regions according to their stage of development and compromising level to achieve the goal. For developing countries that are still not in a mature stage and with limited resources, the priority is to implement a program to achieve the reduction of traditional dependence on others that are usually far away. No auto efficiency can be achieved by the farmers if the multiple obstacles cannot be resolved progressive and collectively.

As a result of the many reasons mentioned, greenhouses farming will be more popular to minimise the risk of losses. While streaming statistic information is determinant for farmers to synchronise with the market's demand, technological applications can also optimise their profits. The participation of different entities to refine agricultural sciences, with no doubt, is a meaningful measurement to improve the efficiency of farming. The sharing of cross border information can help to minimise repeating expensive mistakes.

To improve the financial input, farmlands should not be limited only for traditional food production but diversifying for agribusiness and other economic generations. The concerns demonstrated by diverse international organizations to offer low-rate interests for farming, will surely help the leverage. The indispensability of a more flexible land titling policy will surely help farmers to have expeditious access to financial supports from the banks to strengthen the required cashflow accordingly.

The formation of cooperatives among the farmers' communities and the strengthening of the current ones will provide a stronger negotiating power to achieve better conditions based on a long term partnership, and bind to just causes.

With a collective goodwill, proper implementation and timely adjustments, prelude will be within our sight to comply the mission of feeding the entire world's growing population, without excluding those who are from unprivileged countries. Without the decisive responsibility of the governments to improve the availability of infrastructures, water and energy accessibility, no efforts will ever be enough. Both energies and hope will be drained all the way down to where the underground water is running out.

6. Let Us Help Farming

People do not have to be in the farmland to help farming, but be conscious enough to help balancing the production and consumption, without wasting food.

Furthermore, an increase consciousness of the consumers that ends up with a progressive change in diet, as well as the awareness of food consumption with a diet less dependent on meats, but a greener diet composed of vegetables and fruits, surely helps to achieve a better balance. Unless we manage to change the situation timely, we will be facing a compromising situation with the food production.

Differently to most of other professions, farmers are not guaranteed with a constant income, but only occasionally upon harvests, if lucky enough. In most of the cases, no contracts are secured, and for sure, no monthly payments are promised to have their produces sold at the expected prices, which is a permanent illusion.

By practicing urban farming, people in the cities can enjoy more organic farming by carrying out in small spaces especially for institutional, community and self-consumption. Young generation can have a better identification with faming, provided that basic concepts are introduced as early as possible.

It is quite a dilemma that despite of the unnumberable challenges, there are still people animated to reinforce our mission. Farming is a commitment of generational relay, that bridges up with the daily nature. Alike many other activities, farming demands megabytes of sacrifices, but it is rewarding to be engaged to find the joy.

The task to solve the diverse challenges to produce food, is not limited to the farmers, but a jointly liability. We should head forward for better days, since poverty should not be a heritage for anyone, and in no way for those who produce food to feed the rest of the entire world. Natural resources do run out, but hope does not. Together, and for sure, we can achieve the common goal.

https://doi.org/10.1142/9789811276736_0011

Chapter 11

Co-benefits of Sustainable Food Production and Consumption in Mitigating Air Pollution and Climate Change via the Nitrogen Cycle

Amos P. K. Tai*,†

**Earth and Environmental Sciences Programme, Faculty of Science, The Chinese University of Hong Kong, Hong Kong, China*
†*State Key Laboratory of Agrobiotechnology, and Institute of Environment, Energy and Sustainability, The Chinese University of Hong Kong, Hong Kong, China*
amostai@cuhk.edu.hk

1. Introduction

Food security is one of the most pressing issues affecting humankind. The world population is expected to exceed 9 billion by 2050 (UN, 2015), which, together with rising income and a widespread dietary shift toward heavier meat and dairy consumption in many developing countries, will significantly increase the global food demand (FAO, 2018). It is crucial to enhance food production to cope with the rising demand, but the associated environmental costs can be substantial. Agricultural production is a major driver of greenhouse gas (GHG) emissions, land and freshwater degradation, and biodiversity loss. It contributes significantly to the global emissions of reactive nitrogen (N_r) compounds including nitrous oxide (N_2O), a major GHG, ammonia (NH_3), and nitrogen oxide (NO_x). All these play important roles in the formation of various air pollutants. Therefore, the ways through which food is produced, transported, processed, consumed, and oftentimes-wasted can all be addressed to reduce the climatic and environmental impacts of agriculture. In this chapter, we aim first to summarise the impacts of agricultural production on the atmospheric environment via the nitrogen cascade, and then to examine several cases whereby such impacts may be alleviated via both the supply side (e.g., improving farming methods) and demand side (e.g., adjusting food consumers' dietary habits).

1.1. *Nitrogen cascade*

Human contribution of N_r to the environment has increased more than tenfold since 1860, driven mostly by the substantial growth in both food and energy production. The subsequent fates of N_r and their multiple effects on ecosystems and human health are called the nitrogen cascade (Galloway *et al.*, 2008). Human activities currently contribute 210 Tg-N of N_r to terrestrial and marine ecosystems, which is roughly half of the global nitrogen fixation, i.e., the transformation of inert nitrogen gas (N_2) gas into N_r (Fowler *et al.*, 2013). The majority of anthropogenic sources are associated with agriculture, mostly through the production and excessive use of chemical fertiliser, manure and other animal wastes from livestock, and cultivation-induced nitrogen fixation from fodder legumes. Fossil fuel combustion, which converts N_2 to NO_x under high temperatures, represents another important anthropogenic N_r source. In the natural world, biological nitrogen fixation (BNF) by cyanobacteria and bacteria in symbiotic relationships with legume plants represents the major bulk of natural nitrogen fixation; the rest is mostly associated with high-temperature NO_x production under lightning.

Fertilisers used worldwide commonly contain ammonium (NH_4^+) salt. As organic nitrogen contained in animal manure and excretion is broken down by microorganisms in the process of ammonification, NH_4^+ is also released. These agriculturally derived NH_4^+ ions can volatilise and enter the atmosphere as NH_3 gas, representing the major source of atmospheric NH_3. Various chemolithotrophic bacteria (i.e., metabolising inorganic substrates to obtain energy directly) can also convert NH_4^+ to nitrate (NO_3^-) in the process of nitrification. NO_3^- is highly mobile in soils, and can leach into rivers, lakes and aquifers, causing major water pollution and eutrophication problems. Under anoxic conditions, certain bacteria can conduct anaerobic respiration to obtain energy using NO_3^- as the oxidant instead of oxygen (O_2). The ultimate product of this process, known as denitrification, is N_2, which is released to the atmosphere, completing the nitrogen cycle.

During nitrification and denitrification, significant amounts of N_2O (a potent GHG) and NO_x are released into the atmosphere as byproducts. N_2O is a potent GHG that contributes substantially to global warming; it is relatively long-lived with an atmospheric lifetime of >100 years. In contrast, NH_3 and NO_x relatively quickly react in the atmosphere, contributing to various other N_r compounds; the subsequent deposition of N_r onto the land surface has been shown to lead to various effects on terrestrial ecosystems. These effects include not only faster nitrification, more serious nutrient leaching, soil acidification, and greater N_r losses from soil as NO_x and N_2O (Guo *et al.*, 2010; Lu *et al.*, 2011), but also enhanced plant growth and soil carbon storage in forests and grasslands, especially where nitrogen is a limiting nutrient (Liu *et al.*, 2021; Thomas *et al*, 2010; Zhao *et al.*, 2017). However, such an enhanced growth generally favours the more competitive plant species and may ultimately reduce species diversity of plant communities (Bobbink *et al.*, 2010). The effects of agricultural N_r emissions on air pollution, climate and ecosystems via the nitrogen cascade are summarised in Figure 1.

1.2. *Impacts of agricultural N_r emissions on air pollution and climate change*

The N_r emitted during agricultural activities can subsequently contribute to various air pollutants, including ozone (O_3) and fine particulate matter ($PM_{2.5}$, i.e., particles with a diameter of 2.5 μm or less). Volatile, alkaline NH_3 released from farms and animal operations can react with acidic

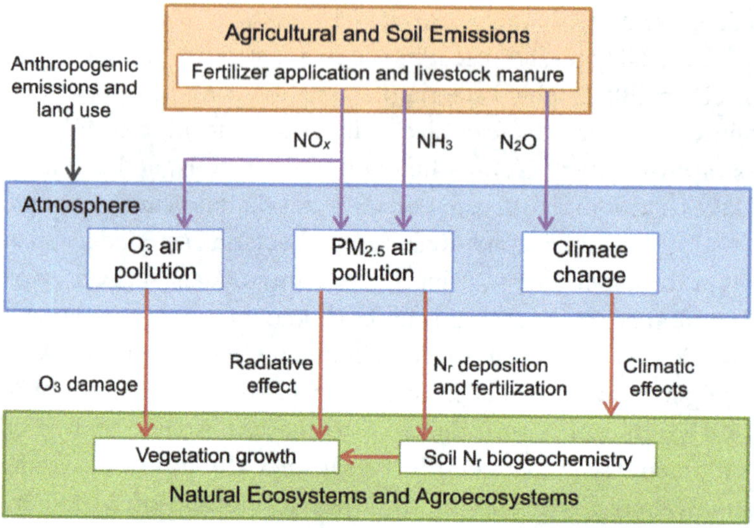

Figure 1. Environmental impacts of agricultural and soil reactive nitrogen (N_r) emissions via the nitrogen cascade, with a focus on atmospheric pathways. The purple arrows represent the direct effects of N_r emissions on the atmosphere by contributing to air pollution and climate change. The red arrows refer to the downstream effects on terrestrial ecosystems via affecting soil biogeochemistry and vegetation growth. The interactions between ozone (O_3), fine particulate matter ($PM_{2.5}$) and climate within the atmosphere are implicitly included.

pollutants (e.g., oxidation products of anthropogenic NO_x and SO_2) to form sulfate-nitrate-ammonium (SNA) particles, which are important constituents of particulate $PM_{2.5}$. High $PM_{2.5}$ concentration in turn causes haze and smog that lower local visibility and severely harm human health, claiming ~3 million premature deaths every year worldwide (WHO, 2016). Due to the relatively low solubility of HNO_3 in aerosol water, the formation of NO_3^- particles is strongly sensitive to the relative abundance of NH_3 and H_2SO_4. Agricultural NH_3 thus plays an important role in determining the atmospheric concentration of SNA particles. Indeed, the total air pollution-related health damage arising from N_r emissions in China is estimated to amount to US$19–62 billion in 2008, accounting for 0.4–1.4% of China's gross domestic product, of which more than half is attributable to NH_3 and the rest mostly to NO_x (Gu *et al.*, 2012).

Meanwhile, NO_x is an important precursor for tropospheric O_3 formation via the photochemical reactions with carbon monoxide (CO), methane (CH_4) and volatile organic compounds (VOCs). In regions where air pollution is a major concern, the contribution to NO_x from agriculture is usually dwarfed by that from fossil fuel combustion. However, anticipating a general downward trend in anthropogenic NO_x emissions following strict regulatory efforts worldwide including in China, agricultural and soil NO_x is expected to be increasingly important to regional $PM_{2.5}$ and O_3 pollution in the future. For instance, a study has proposed that a 50% reduction in NH_3 emission nationally via improving agricultural management, along with a targeted emission reduction (15%) for SO_2 and NO_x, can alleviate $PM_{2.5}$ pollution by 10–20% by suppressing SNA formation (Liu *et al.*, 2019). N_r deposition would decrease by 30–40%. However, this NH_3 reduction would aggravate acid rain, with a decrease of as much as 1.0 unit in rainfall pH and substantial increase in areas suffering heavy acid rain.

Agricultural N_r also contributes substantially to climate change. According to a special report of the Intergovernmental Panel on Climate Change, the global food systems contribute to a global total GHG emission of 11–19 Gt CO_2-equivalent (CO_2e) per year, corresponding to 21% to 37% of overall anthropogenic emissions (Mbow *et al.*, 2019). A recent study estimated the total global GHG emission from the food systems to be 18 Gt CO_2e yr^{-1}, among which about 10% is attributable to agricultural N_2O emissions, 35% to methane (CH_4) emissions, 52% to CO_2 emissions, and 2% to fluorinated gases (Crippa *et al.*, 2021). The N_2O emitted from the food systems is mostly a by-product of nitrification and denitrification, which are ultimately driven by the substantial amount of nitrogenous fertiliser used in croplands and animal excretions and wastes from livestock industries. While N_2O is a climate-warming GHG, the SNA particles formed from anthropogenic N_r from both fossil fuel combustion and agriculture as described above can have the opposite, cooling effect by scattering solar radiation, reflecting it back to space, and enhancing cloud formation. Such a cooling effect may partly or nearly completely offset the warming effect of agricultural N_2O (Pinder *et al.*, 2012), but if only particles associated with agricultural N_r alone is considered, the net effect of agricultural N_r per se is still an overall warming.

1.3. *Focusing on Chinese food production and consumption*

China is an archetypal case of the complex issues behind its substantial yet still rapidly rising food consumption and production. It is the most populous country in the world as of 2020. Per capita meat consumption in China has been increasing rapidly since the 1980s (roughly fivefold by 2010s), rendering China the number-one meat consumer in the world now (He *et al.*, 2018; Liu *et al.*, 2021). It is the largest N_2O emitter in the world (~1.0 Tg-N of N_2O emitted annually), accounting for 5–10% of the global total emission in the recent two decades (Gu *et al.*, 2012; Tian *et al.*, 2011). It is also the largest emitter of agricultural NH_3 (8–12 Tg-N of NH_3 released annually), contributing to 20–30% of the global total emission (Gu *et al.*, 2012; Liu *et al.*, 2021; Zhang *et al.*, 2018). Meanwhile, ambient $PM_{2.5}$ is responsible for 1.1–2.0 million deaths in China in 2017, ranked as the number-four leading cause of mortality (Zhou *et al.*, 2019). Earlier studies estimated that the Chinese government needs to spend up to US$3.3 billion to compensate for the health impacts of 1 Tg-N of emitted NH_3, reflecting the costs of hospital care and rehabilitation from the associated $PM_{2.5}$ pollution (Gu *et al.*, 2014). The future trend in Chinese agricultural production will continue to have tremendous impacts on human and environmental health by affecting air pollution. It is critically important to examine how various socioeconomic paths under different assumptions of dietary options, land use and agricultural technology may influence N_r emissions, air quality and health, so that policy making and public choices can be informed accordingly.

Agricultural production to feed a growing world population and worldwide changes in dietary patterns have had substantial impacts on human and environmental health. A more nutritionally enriched diet benefits human health, but such benefits are increasingly offset by the disease burden coming from more meat-intensive diets and overnutrition in many countries. Per capita meat consumption in China has been increasing rapidly since the 1980s, which has been shown to impose increasing risks of cardiovascular diseases, cancers and type II diabetes to the population; now dietary risks are the number-one leading cause of mortality in China, estimated to contribute to 3.1 million premature deaths in 2017 (He *et al.*, 2018; Zhou *et al.*, 2019). Meanwhile, environmental degradation incurred by increasing and changing patterns of food production may add further burdens to human health. Agricultural intensification and industrialisation have already rendered China among the most polluted countries by the early 2010s, when the annual mean population-weighted exposure to $PM_{2.5}$ is on average ~60 μg m^{-3} for China and >100 μg m^{-3} in some eastern regions (Shaddick *et al.*, 2018). Dietary changes, along with population growth, are key drivers of changes in food production, and thus pose a threat to public health not only directly via malnutrition but also indirectly via degrading air quality and respiratory health. Such an "indirect" health cost of dietary changes for China has only recently been quantified (Liu *et al.*, 2021), as will be discussed later in this chapter. In view of the above, addressing agricultural food production and consumption as a gateway toward simultaneously maintaining environmental quality and human health is even more pressing in China and other rapidly developing regions than other parts of the world.

2. Supply-side Mitigation of N_r Pollution via Improving Food Production Methods

On the production side, more cost-effective agricultural N_r management has been shown to hold large potential to reduce the substantial agricultural N_r leakages and the subsequent climatic, ecological, and human health impacts, while securing the food supply. One recent study has attempted to quantify such potential for China (Guo *et al.*, 2020). The authors recognised that currently, Chinese farmers tend to apply excessive N_r fertiliser (e.g., at levels 50–150% higher than their counterparts in the US), and more than half of the fertiliser N_r is lost to the environment, reflecting relatively low nitrogen use efficiency (NUE) in croplands across the country (Chen *et al.*, 2021). The farmers generally have little practical knowledge or incentive to practice the recommended 4R principles for nutrient management (i.e., the right time, right amount, right form, and right method), at least in part due to the heavy government subsidies on fertiliser purchase. The authors conducted an integrated assessment of four particular recommended strategies to improve agricultural N_r management: (1) improved farm management practices (including moderate farm management improvements and reductions in N_r fertiliser application rates); (2) precision farming with machine deep placement of fertilisers; (3) enhanced-efficiency fertiliser use (including the use of controlled-release fertilisers and urease inhibitors); and (4) improved manure management (including manure acidification, aerobic composting, cropland injection and improved animal feed). These management scenarios, when implemented separately following recommendations derived from Chinese field experiments, can reduce annual NH_3 emission from croplands by 6–11%, but when combined they can lead to an overall reduction by up to 34%.

Using a process-based NH_3 emission model combined with an atmospheric chemical transport model (CTM) and subsequent health impact analysis, Guo *et al.* (2020) thus found that simultaneous implementation of the four strategies above provides the largest benefits, which include reductions in annual mean $PM_{2.5}$ concentrations by up to 8 μg m^{-3} (28%) and the associated avoidance of ~30,500 premature deaths per year. They also found increases in grain yields and grain N_r use efficiency, reductions in NO_3^- leaching and runoff, and reduced GHG emissions. The total benefits can amount to US$30 billion per year, which exceed the US$18 billion per year in costs.

There are various ways through which N_r fertiliser rates can be reduced. One of them not addressed specifically by the study above is intercropping, which has been heralded as a sustainable agricultural practice that can reduce the environmental impacts of agriculture, but its potential benefits beyond the farm scale have rarely been examined. Intercropping entails two or more crops growing in the same field side by side in the same or overlapping growing season. It generally improves land-use efficiency because less land is required to yield the same quantities of monoculture crop grains. The most beneficial way is intercropping non-legume crops with legumes, which can facilitate BNF, make the excess N_r fixed accessible to the other crops, enhance NUE and thus be able to maintain the same yields of non-legume crops with less fertiliser input than their monoculture counterparts (Yong *et al.*, 2015).

In our recent study, we thus investigated the environmental and yield benefits of intercropping in China (Fung *et al.*, 2019). A model for below-ground mutualistic interactions between intercropped crops was developed and used to simulate and quantify the benefits of nationwide adoption of maize-soybean systems in China in terms of enhanced crop production, reduced fertiliser application and lower NH_3 emission. We further examined how such a decrease in NH_3 emission could lessen the downwind formation of $PM_{2.5}$ using a CTM, and showed that annual mean $PM_{2.5}$ concentrations can decrease by up to 1.5 μg m^{-3}, leading to an annual net economic benefit of US$67 billion, of which US$13 billion arises from saved health costs from reduced air pollution, US$59 billion from gains in grain production, US$0.47 billion from saved costs of fertilisers, offset finally by US$5.9 billion from additional labour and machinery costs incurred by switching from monoculture to intercropping (Figure 2).

The above two studies demonstrate the economic, environmental and ecological values of improving agricultural N_r management (including intercropping as one of the various approaches) in simultaneously promoting food security and public health while addressing multiple environmental and climatic challenges. They serve as a basis for policy consideration as governments, farmers and other stakeholders explore more sustainable farming options and approaches. These points have been highlighted by a recent review on the topic (Balasubramanian *et al.*, 2021).

3. Demand-side Mitigation of N_r Pollution via Adjusting Food Consumption Patterns

Agricultural production is ultimately driven by food demands from consumers. It is possible to attribute the air quality-related health damages arising from food production to specific food consumption and dietary choices made by the populations. A recent study has done exactly that, focusing on the US (Domingo *et al.*, 2021). The authors integrated crop

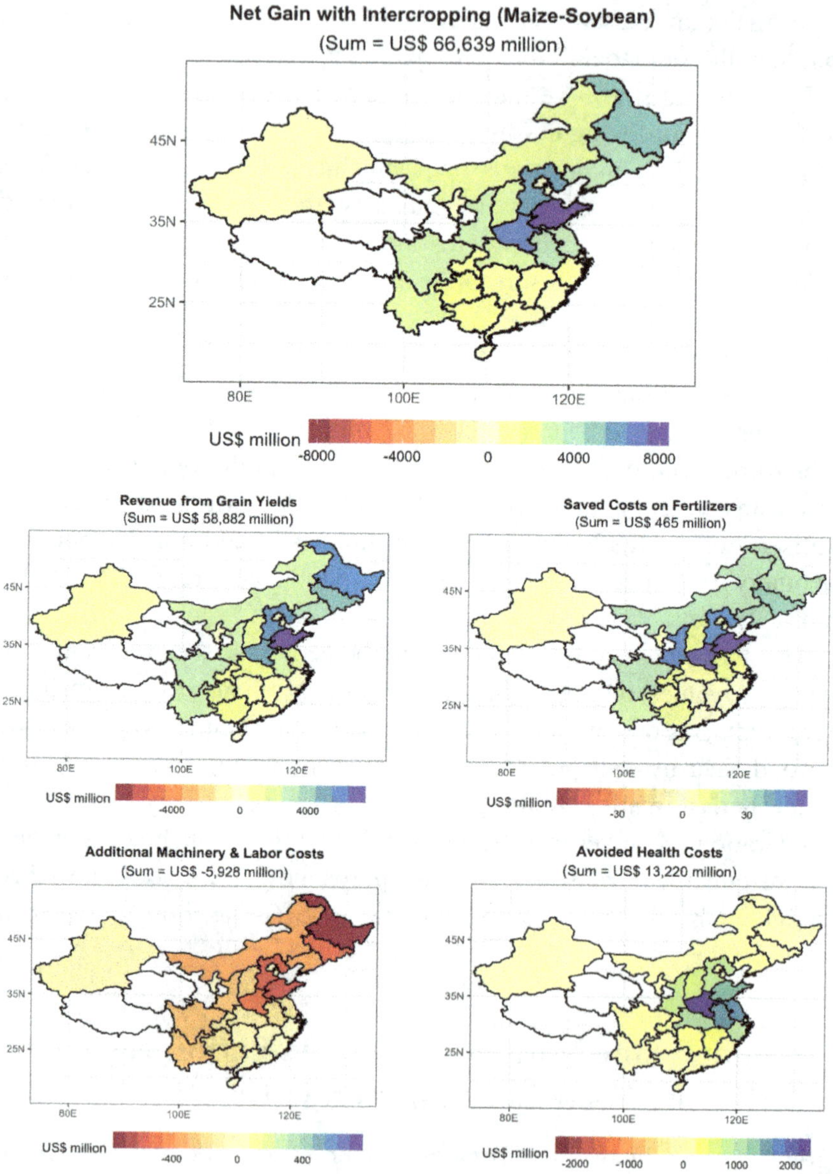

Figure 2. Cost-benefit analysis for the nationwide adoption of the maize-soybean intercropping system in continental China (excluding Taiwan and other outlying islands). Net benefits are calculated by subtracting labour and machinery costs (which take up ~5/6 and ~1/6 of the additional costs, respectively), urea fertiliser expenses, and health costs associated with $PM_{2.5}$ pollution from the total revenues from grain sales. Adopted from Fung *et al.* (2019).

and livestock type-specific air pollutant emission inventories with a series of regional CTMs, sensitivity analysis and health impact analysis to find that agricultural production in the US leads to 17,900 air quality-related premature deaths per year, most of which are driven by NH_3 emissions from livestock waste and fertiliser application, consistent with what was found for China as described above. Furthermore, 80% of those deaths are attributable to animal-based foods, both directly from animal production and indirectly from the cultivation of feed crops for animals. They suggested that dietary shifts toward more plant-based foods can reduce those air quality-related mortalities arising from agricultural production by 68–83%, a substantial health benefit of adjusting dietary habits that is beyond the immediate health benefits for the individual food consumers.

Our recent study has examined the same issue for China (Liu *et al.*, 2021). Figure 3 demonstrates the overall idea. China's population has been persistently rising in the recent decades, having increased by nearly 40% between early 1980s and 2010s. There is also a fivefold increase in per capita daily consumption of meat among the Chinese population in the same period. Both factors together have substantially increased

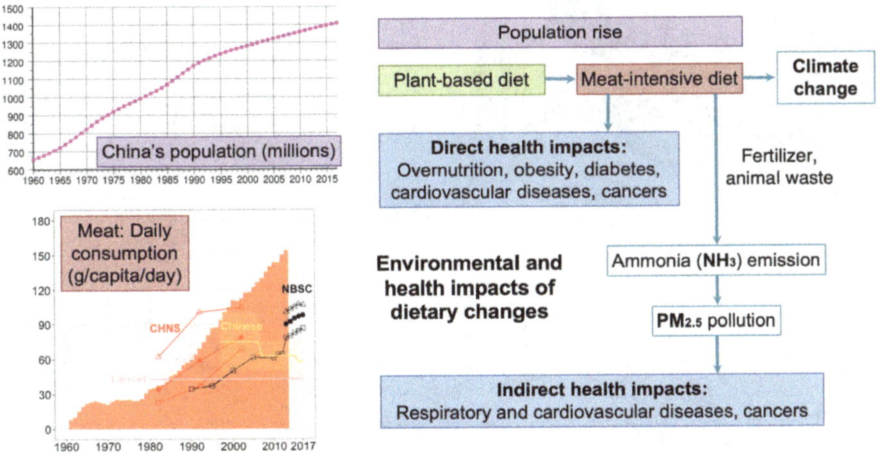

Figure 3. Thematic diagram showing how dietary changes in the form of higher meat consumption can not only directly impact human health by increasing the risks of diet-related diseases and worsen climate change via GHG emissions, but also worsen air pollution, leading to unexpected health consequences and premature deaths. Adopted from Liu *et al.* (2021).

agricultural production in China, leading to more GHG emissions that contribute to climate change, while higher meat consumption has also led to greater direct health risks of various diet-related diseases for the individual food consumers. What is less known, however, is how the resulting enhanced NH_3 emission and $PM_{2.5}$ pollution may lead to indirect health risks of air pollution-related diseases.

We therefore combined a food type-specific NH_3 emission model with a CTM, sensitivity experiments and health impact analysis to estimate the $PM_{2.5}$-related health impacts arising from different food types. We found that changes in the Chinese diet over 1980–2010, mainly in the form of higher meat consumption, have increased NH_3 emissions from the agricultural sector by 63% and increased annual mean $PM_{2.5}$ concentration by up to 10 μg m^{-3}, which is ~20% of the total increase over the same period from all sources and leads to ~91,000 more air pollution-related premature deaths every year (Figure 4). Such effects are more than double that driven by increased food production solely due to population growth. It is also noteworthy that whereas the rising meat demand

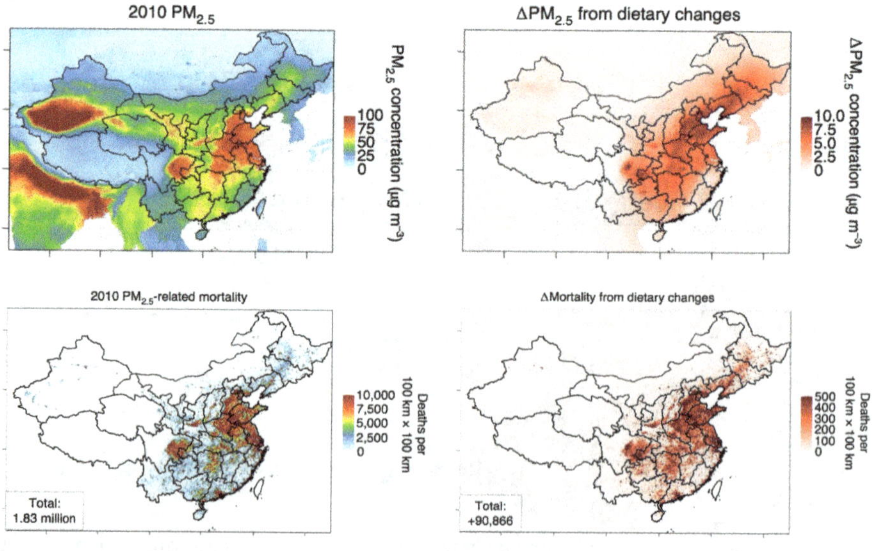

Figure 4. Worsening of $PM_{2.5}$ air quality due to dietary changes (including demands for meat, feed crops and food crops), and the associated indirect health costs in terms of premature mortalities. Adopted from Liu *et al.* (2021).

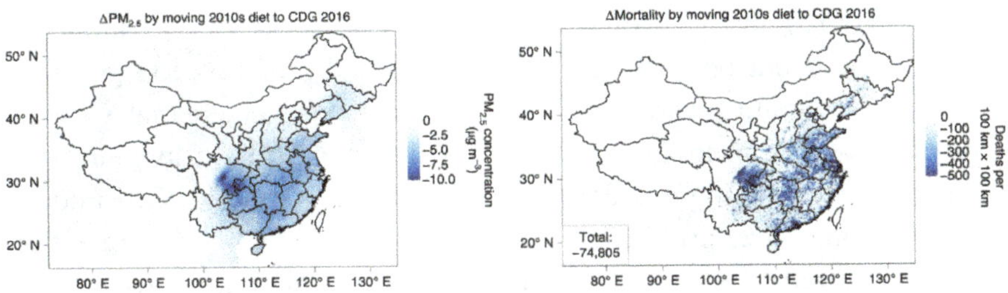

(a) (b)

Figure 5. Potential environmental and indirect health benefits of moving from the current 2010s diet to the healthier, less meat-intensive diet recommended by the 2016 Chinese Dietary Guideline in terms of changes in annual mean particulate matter concentration and pollution-related premature mortality. Adopted from Liu *et al.* (2021).

is driven mainly by the richer coastal city populations, the majority of the associated air pollution-related health burden falls onto the poorer populations in the agricultural regions who consume less meat. This reflects the health inequity associated with meat consumption in China, an environmental justice dimension that deserves attention from a public policy perspective.

We further found that shifting the current meat-intensive diet toward a less meat-intensive one according to the 2016 Chinese Dietary Guideline can decrease NH_3 emission by ~20% and $PM_{2.5}$ by up to 6 μg m^{-3}, and avoid ~75,000 annual premature deaths related to air pollution in China (Figure 5). We therefore suggested that nationwide adoption of a more plant-based diet could serve as a strategy to mitigate the serious air pollution problems in China, consistent with the suggestion by Domingo *et al.* (2021) for the US.

4. Concluding Remarks

We expect food consumption patterns, dietary choices, and production to meet the corresponding food demands will continue to shape air pollution and GHG emissions in China, especially as anthropogenic emissions from fossil fuel combustion are on the decline (Liu *et al.*, 2019). The future food demands will be primarily shaped by three factors: (1) population growth, (2) per capita total food consumption (e.g., daily dietary calorie intake), and (3) dietary composition. The latter two factors are both strongly dependent on income level and personal preference (e.g., taste, convenience, health awareness). United Nations projected that China's population will peak in around 2030, reaching 1.42 billion in year 2030 and thereafter declining to 1.35 billion in year 2050 (UN, 2015). Building upon the population projection, the Food and Agriculture Organization further projected future food consumption and production trends country by country based on three scenarios: "stratified societies", "business as usual", and "towards sustainability" (FAO, 2018). Depending on the scenario, China's dietary energy consumption per capita is projected to increase by 4–8% in 2030 and 2–6% in 2050, from year-2012 level of 2971 kcal per person per day. Meat product consumption per capita will rise by 9–16% in 2030 and 5–14% in 2050, from year 2012-level of 632 kcal per person per day. To satisfy the growing demands, total crop harvested area in China has to expand by 9–13% in 2030 and 3–9% in 2050, and animal herd size has to increase by 1–18% in 2030 and change by –5% to +9% in 2050.s

The food trends as summarised above are bound to have profound impacts on Chinese air quality and health as well as GHG emissions via the nitrogen cascade, the extent of which will strongly depend on the path taken by Chinese policy makers, agricultural food producers and consumers at large. Given the currently low NUE across all Chinese croplands, farmers have tremendous untapped potential to improve agricultural N_r management on the fields. In addition, adjustments in dietary choices alone, such as by adopting a healthier, less meat-intensive diet, or even a vegetarian or vegan diet, can have considerable impacts on production and emissions. If the populations do not adjust their diets, agricultural and food production technology may have to be improved much beyond the current state to offset the additional impacts of increased food consumption. The studies reviewed above have demonstrated how agricultural food producers and consumers can together help address the problems, reduce the impacts of the agricultural sector while ensuring a sustainable food supply, and prevent the direst environmental and climatic consequences.

References

Balasubramanian, S. *et al.*, 2021. The food we eat, the air we breathe: A review of the fine particulate matter-induced air quality health impacts of the global food system. *Environ. Res. Lett.*, 16(10). doi:ARTN 10300410.1088/1748-9326/ac065f

Bobbink, R. *et al.*, 2010. Global assessment of nitrogen deposition effects on terrestrial plant diversity: A synthesis. *Ecol. Appl.*, 20(1): 30–59. doi:Doi 10.1890/08-1140.1

Chen, F. J. *et al.*, 2021. Breeding for high-yield and nitrogen use efficiency in maize: Lessons from comparison between Chinese and US cultivars. *Adv. Agron.*, 166(166): 251–275. doi:10.1016/bs.agron.2020.10.005

Crippa, M., Solazzo, E., Guizzardi, D., Monforti-Ferrario, F., Tubiello, F. N., and Leip, A., 2021. Food systems are responsible for a third of global anthropogenic GHG emissions. *Nat. Food*, 2(3): 198–209. doi:10.1038/s43016-021-00225-9

Domingo, N. G. G. *et al.*, 2021. Air quality-related health damages of food. *Proc. Natl. Acad. Sci. USA*, 118(20). doi:ARTN e201363711810.1073/pnas.2013637118

FAO, 2018. *The future of food and agriculture — Alternative pathways to 2050.* Retrieved from Rome, Italy: www.fao.org/publications

Fowler, D., 2013. The global nitrogen cycle in the twenty-first century. *Philos. Trans. R. Soc. B Biol. Sci.*, 368(1621): 20130164. doi:ARTN 2013016410.1098/rstb.2013.0164

Fung, K.M., Tai, A.P.K., Yong, T.W., Liu, X.M., and Lam, H.M., 2019. Co-benefits of intercropping as a sustainable farming method for safeguarding both food security and air quality. *Environ. Res. Lett.*, 14(4): 044011. doi:ARTN 04401110.1088/1748-9326/aafc8b

Galloway, J.N., 2008. Transformation of the nitrogen cycle: Recent trends, questions, and potential solutions. *Science*, 320(5878): 889–892. doi:10.1126/science.1136674

Gu, B.J. *et al.*, 2012. Atmospheric reactive nitrogen in China: Sources, recent trends, and damage costs. *Environ. Sci. Technol.*, 46(17): 9420–9427. doi:10.1021/es301446g

Gu, B.J., Sutton, M.A., Chang, S.X., Ge, Y., and Chang, J., 2014. Agricultural ammonia emissions contribute to China's urban air pollution. *Front. Ecol. Environ.*, 12(5): 265–266. doi:10.1890/14.Wb.007

Guo, J.H. *et al.*, 2010. Significant acidification in major Chinese croplands. *Science*, 327(5968): 1008–1010. doi:10.1126/science.1182570

Guo, Y.X. *et al.*, 2020. Air quality, nitrogen use efficiency and food security in China are improved by cost-effective agricultural nitrogen management. *Nat. Food*, 1(10): 648–658. doi:10.1038/s43016-020-00162-z

He, P., Baiocchi, G., Hubacek, K., Feng, K.S., and Yu, Y., 2018. The environmental impacts of rapidly changing diets and their nutritional quality in China. *Nat. Sustain.*, 1(3): 122–127. doi:10.1038/s41893-018-0035-y

Liu, M.X. *et al.*, 2019. Ammonia emission control in China would mitigate haze pollution and nitrogen deposition, but worsen acid rain. *Proc. Natl. Acad. Sci.USA*, 116(16): 7760–7765. doi:10.1073/pnas.1814880116

Liu, X.Y., Tai, A.P.K., Chen, Y.F., Zhang, L., Shaddick, G., Yan, X.Y., and Lam, H.M., 2021. Dietary shifts can reduce premature deaths related to particulate matter pollution in China. *Nat. Food*, 2(12): 997–1004. doi:10.1038/s43016-021-00430-6

Liu, X.Y., Tai, A.P.K. and Fung, K.M., 2021. Responses of surface ozone to future agricultural ammonia emissions and subsequent nitrogen deposition through terrestrial ecosystem changes. *Atmos. Chem. Phys.*, 21(23): 17743–17758. doi:10.5194/acp-21-17743-2021

Lu, M. *et al.*, 2011. Responses of ecosystem nitrogen cycle to nitrogen addition: A meta-analysis. *New Phytol.*, 189(4): 1040–1050. doi:10.1111/j.1469-8137.2010.03563.x

Mbow, C. *et al.*, 2019. *Food Security in Climate Change and Land: An IPCC Special Report on Climate Change, Desertification, Land Degradation, Sustainable Land Management, Food Security, and Greenhouse Gas Fluxes in Terrestrial Ecosystems.* Retrieved from

Pinder, R.W., Davidson, E.A., Goodale, C.L., Greaver, T.L., Herrick, J.D., and Liu, L.L., 2012. Climate change impacts of US reactive nitrogen. *Proc. Natl. Acad. Sci. USA*, 109(20): 7671–7675. doi:10.1073/pnas.1114243109

Shaddick, G. *et al.*, 2018. Data integration for the assessment of population exposure to ambient air pollution for global burden of disease assessment. *Environ. Sci. Technol.*, 52(16): 9069–9078. doi:10.1021/acs.est.8b02864

Thomas, R.Q., Canham, C.D., Weathers, K.C., and Goodale, C.L., 2010. Increased tree carbon storage in response to nitrogen deposition in the US. *Nat. Geosci.*, 3(1): 13–17. doi:10.1038/Ngeo721

Tian, H. Q. *et al.*, 2011. Net exchanges of CO_2, CH_4, and N_2O between China's terrestrial ecosystems and the atmosphere and their contributions to global climate warming. *J. Geophys. Res. Biogeosci.*, 116. doi:Artn G0201110.1029/2010jg001393

UN, 2015. *World Population Prospects: The 2015 Revision, Key Findings and Advance Tables*. Retrieved from

WHO, 2016. *Ambient Air Pollution: A Global Assessment of Exposure and Burden of Disease*. Retrieved from Geneva.

Yong, T. W. *et al.*, 2015. Characteristics of nitrogen uptake, use and transfer in a wheat-maize-soybean relay intercropping system. *Plant Prod. Sci.*, 18(3): 388–397. doi:DOI 10.1626/pps.18.388

Zhang, L. *et al.*, 2018. Agricultural ammonia emissions in China: Reconciling bottom-up and top-down estimates. *Atmos. Chem. Phys.*, 18(1): 339–355. doi:10.5194/acp-18-339-2018

Zhao, Y.H., Zhang, L., Tai, A.P.K., Chen, Y.F., and Pan, Y.P., 2017. Responses of surface ozone air quality to anthropogenic nitrogen deposition in the Northern Hemisphere. *Atmos. Chem. Phys.*, 17(16): 9781–9796. doi:10.5194/acp-17-9781-2017

Zhou, M. G. *et al.*, 2019. Mortality, morbidity, and risk factors in China and its provinces, 1990–2017: A systematic analysis for the Global Burden of Disease Study 2017. *Lancet*, 394(10204): 1145–1158. doi:10.1016/S0140-6736(19)30427-1

https://doi.org/10.1142/9789811276736_0012

Chapter 12

My Experience as a Development Worker: Working on Food Security and Nutrition in Africa

Steven Chun-Kit Li

Former Programme and Policy officer,
United Nations World Food Programme
stephenlkkpm6a31@yahoo.com.hk

Unlike in other developed countries or regions, very few young people in Hong Kong are engaged in the humanitarian and development aid sector. To foster and nurture talents of the sector, universities and institutions in the west offer specific courses related to humanitarian aid and development assistance. Organizations working on these two areas, namely NGOs, regional and international organizations (such as the European Union and the United Nations), national development agencies which support the assistance of their countries for targeted developing countries offer job opportunities for those passionate youth. Hong Kong is a complete contrast. Humanitarian relief and development assistance are never on the job list. Both teachers and school students applying for local universities do not have the faintest notion of development sector as academic programmes on sustainable development, organizations on development aid and relevant job opportunities are all thin on the ground. I was also no exception, and I could not imagine being devoted to the relevant sector.

Some experiences I had during my late secondary and tertiary education enabled me to be better prepared for my later career development. Raised in a grassroot family in Hong Kong, I had a tough time at school. I had no motivation to study as I deemed the subjects pointless and useless. Language teaching style at my time — cramming grammar rules, doing drilling exercises and reciting texts for dictation — was dry and boring. It was a discouraging and dreadful experience. At the end of the day, I got poor grades in a public exam, similar to GCSE. I failed in English, which barred me from matriculation classes.

A one-year exchange programme which I learnt about in English lessons during my fifth year of secondary education became an option and eventually metempsychosis in which I found my path to reach my purpose, and my calling in life. English-speaking countries are always the most popular destinations and mine with the hope to improve English but my exam grades limited my options to Norway, Sweden and Hungary. Norway came top in my preference bucket list on my application after a discussion with my English teacher, who gave me two reasons: Norwegians speak good English and Norway is not as cold as Sweden (though my teacher did not know Swedes also speak English well).

The programme fee was much lower than that of studying abroad due to the non-profit making nature of the exchange organization but it was still unaffordable for my family. I must thank my mother for going the extra mile to help make my hopes and dreams — exploring the world at the age of 17 — possible. The year in Norway broadened my horizon. It was the first time I had lived in a rural area. My host family lived on a farm in a small village with a population of 60 with sheep and deer wandering around. In Norway, unlike the agriculture in continental Europe, small and medium-size family-run farms dominate the agricultural sector. Our neighbour next door was a farmer, working on his own farm and farmland rented from my host family.

A lot of food and drink consumed in my host family were also produced locally in Norway. My host father had a lot of apple trees in front of his house. In summer, he harvested apples which were sold to aggregators, juiced or fridged for future consumption. Local produce was on the kitchen table, from apples to cods, and from milk to cheese. In addition, I was surprised when told to drink tap water, which was unimaginable in Hong Kong.

School was another new experience. On the first day at school, I was thrilled to learn that there was no school uniform as a lot of my conflicts with disciplinary teachers stemmed from school uniform, and I could wear almost whatever I liked to attend school. Apart from the freedom of dressing, language learning was also different from the Hong Kong education system. Local students had to choose a language class — French, German or Spanish as a third language while learning English was compulsory. I picked French as only the French class offered elementary level. Language learning in Norway was more interactive and there were more teacher–student and student–student interactions owing to the size of a language class. The ratio was one to ten, while it was one to forty at my class in Hong Kong. There were no mock exams and past papers but real language tasks — discussions. Some of my classmates spoke very fluent English as English was used in daily life — while playing online video games and watching English TV series. It was at that moment I started to realise the purpose of learning English and a third language, which ignited my interest in languages. I continued to learn French after my exchange year in Norway.

What I gained in Norway were valuable assets which have steered me in future studies and career. Starting to learn French in Norway was a key to studying in France as French was and still is the medium of instructions in public universities. I majored in a field of study in social sciences, European studies, which focused on the development in European integration at political, economic and social levels. I learnt about the importance of agriculture in Europe and the introduction and implementation of policies to protect European agriculture. Initiated by France and launched in 1962, the European Union's common agricultural policy (CAP) has been a partnership between agriculture and society, and between Europe and its farmer. According to the European Union (EU),[1] the objectives of CAP are to:

- support farmers and improve agricultural productivity, ensuring a stable supply of affordable food;
- safeguard EU farmers to make a reasonable living;

---

[1] https://agriculture.ec.europa.eu/common-agricultural-policy/cap-overview/cap-glance_en?fbclid=IwAR3C7cKwpQSKKa2NLtr8Upx9S3dBOkb-HEAkj1n8sMh_JKPCI1Iv6XfBtWk

- help tackle climate change and the sustainable management of natural resources;
- maintain rural areas and landscapes across the EU;
- keep the rural economy alive by promoting jobs in farming, agri-food industries and associated sectors.

Apart from lectures, I was able to gain knowledge outside the lecture theatre, and the Paris International Agricultural Show was one. Held in the first trimester each year, its annual exhibition is always a major agricultural event for the sector. In 2019 the exhibition attracted more than 600,000 visitors from all over the world and 400 exhibitors.[2] My first visit to the exhibition was really inspiring, as I discovered various regional agricultural specialties of France. Livestock was also brought to the exhibitions and featured there. Some animals astounded me, like those cows that were indeed taller than me, and some creative activities blew my mind, like pageants for cows, in which producers brought their "sexiest" cows to compete for the title. As a whole, studying in Paris was an amazing experience, as the city offered a lot of opportunities for me to learn different things and experience something which is unusual in my home town.

After my studies in France I left my footprint in Vienna, the capital of Austria with full of interesting history and culture. As I always wanted to study in an ancient university, I pursued a master's degree in development studies in the University of Vienna. The master study adopts an inter- and multi-disciplinary approach to explain the political, economic, social and environmental development in the world, especially in the developing countries. Development studies is a discipline which is not so popular in Hong Kong but in Europe and the US, it is offered in quite a number of universities. After the independence of the former western colonies, this field of study emerged to support the development of the new-born countries by their developed counterparts in terms academic research and relevant talent training. After the first year of the theoretical learning in my master's degree, I applied for an internship to learn hands-on experience

---

[2] https://en.salon-agriculture.com/The-Show/The-history-of-the-show?fbclid=IwAR3aCktDBnvBqXsnlZpVfYrXqsmNPoPaI4yUkxDL65klbe5vkiVB-jlfO9s

in sustainable development in developing countries, a choice very different from my course-mates, because I was keen to observe and study the issues on the ground and have direct contact with local stakeholders. Africa was in my mind. Among several applications, I received a prompt favourable response from the office of the United Nations Industrial Development Organization in Mali, a French-speaking country in West Africa.

The experience was even more challenging and enriching than I had imagined. Although Mali was inflicted by terrorism, political crises and civil wars in recent decades, I accepted the offer and stayed there for three months. Upon arrival, I realised that security in Mali was worse than expected, when I attended a safety briefing at the base of the UN Peacekeeping Mission, including how to react and evacuate when facing terrorists. Fortunately, I was able to return to Vienna safe and sound. This internship, unexpectedly, paved the way for my career in the development sector, by allowing me to better understand the challenges developing countries face every day. During my internship, I was assigned to conduct research on the value chain of shea and gum Arabic. The latter is a material used in food industry, quite unknown to us but widely used as a stabiliser in soft sweets and soft drinks, indicated as E141 on the list of ingredients on food packaging. Mali is one of the main producers of crude gum Arabic. The exportation of crude gum Arabic to developed countries could not make a fortune for its exporters in Mali, while its processing in developed countries added a much higher value. To investigate why the processing was rarely done in developing countries, I had interviews with a few key stakeholders in the sector, including the owner of a company which collected gum Arabic and preliminarily processed the raw materials. After the investigation, I realised that developing countries encountered challenges at different levels. In the first place, the processing involved sophisticated techniques and modern equipment, which required investment in human resources and capital. Nevertheless, with risk-averse commercial banks in West Africa, it was not easy for entrepreneurs to secure a loan, not to mention a low-interest loan, for investments. Interest rates in most African countries reached double digits up to 20%, much higher than that in developed countries. Accumulation of capital was prevalent for further investment. Quality control was another issue. Gum

Arabic used in food products in developed countries must meet the standard of consumer countries. Only certified ingredients were allowed on the market of developed countries. Due to very limited qualified laboratories in Africa, products from Africa had to be sent to specific laboratories in Europe for testing and certification, which inevitably increased the production cost. Even if African producers overcame all the aforementioned obstacles, to gain trust from food manufacturers in developed countries was another hurdle. Food production and imported ingredients in developed countries must comply with very strict requirements of food law, so manufacturers would never take risks when ingredients could be procured from processors of developed countries.

My internship experience made me realise that such situation does not only happen to gum Arabic, but other raw materials. "The curse of raw materials" limits the socio-economic development of many raw material exporting countries, especially those in Africa, preventing them from generating higher income and creating more jobs in relevant sectors.

With this first-hand experience in Mali, I made up my mind to pursue my career in the development sector. In 2019, I accepted a job offer for a position from the United Nations World Food Programme (WFP) in Côte d'Ivoire, which is a French-speaking country. I believe my proficiency in French and English together with my previous experience in West Africa made me stand out from the crowd. Once I arrived in Côte d'Ivoire, the first impression in its capital, Abidjan, is very surprising. On the way from the airport to my apartment, there were skyscrapers, which looked like those in commercial areas in China. In Côte d'Ivoire, there are a number of local languages. People can generally speak fluent French, which means that most of them have been to schools for education. I could also find a few restaurants and shops, which I went to quite often during my study in France. Therefore, It is relatively easy for foreigners to settle down and live in Abidjan.

Côte d'Ivoire is a better off country in West Africa. However, the region is inflicted by food insecurity and malnutrition. My adventure with WFP in West Africa started my career in food security and nutrition, which is a topic rarely discussed in Hong Kong. Brought up in an affluent city with food from all corners of the world, food shortage, food security and nutrition were never on my plate. Unfortunately, millions living on the same planet are hit by unavailability of and inaccessibility (due to high price) to nutritious food. According to the publication "The State of Food Security and Nutrition in the World 2022",[3] between 702 and 828 million people were affected by hunger in 2021. Asia accounts for more than 50% of the affected population, while Africa makes up more than a quarter of the population. Some groups are more impacted by food insecurity and malnutrition than others, such as women. About 31.9 percent of women in the world were moderately or severely facing food insecurity compared to 27.6 percent of men. In recent years, the COVID pandemic and the war between Russia and Ukraine have aggravated hunger in developing countries. The rise in food price limits the access of the population in developing countries. Unlike the population in developed countries, food purchase constitutes a large part of their income. For example, food purchase accounts for up to 40% of the income of the population in sub-Saharan Africa.[4] The surge in food price forces them to spend a larger proportion of their income on food. The affected population may cope with the situation by buying less nutritious food, which can be more expensive. The less diverse diet may lead to the insufficient intake of nutrients, causing long-term effects in child physical and cognitive growth. Learning ability is affected due to low cognitive capacity. The future labour force of the country will thus be less productive, limiting their contribution to economic activities. As a result, the consequences of food insecurity and malnutrition may hinder the socio-economic development of a country.

Sandy deserts, famine and poverty are a few stereotypes of Africa for people who do not know much about Africa. Stereotypes are preconceived notions and sadly many stereotypes are rooted in prejudice — poverty in Africa is due to laziness of the African population, and hunger is due to limited agricultural land. One of many problems with any stereotype is that even if it is true in some cases; it is certainly not true in all cases. I have met a lot of people in Mali and Côte d'Ivoire who work very hard to earn their living: hawking in the middle of the roads, waking up at 5 am to work in farms, driving a taxi from 6 am to 9 pm. Some even work

[3] FAO, IFAD, UNICEF, WFP and WHO. 2022. *The State of Food Security and Nutrition in the World 2022. Repurposing food and agricultural policies to make healthy diets more affordable*. Rome, FAO.

[4] https://www.imf.org/fr/News/Articles/2022/04/28/blog-africa-faces-new-shock-as-war-raises-food-fuel-costs

longer hours to make both ends meet, sometimes under terrible conditions. Regarding the availability of agricultural land, Africa is not less favoured than other continents. Although it has the third largest desert in the world — the Sahara Desert, an extensive area of fertile soil, for example in the Democratic Republic of Congo (DRC) is not lacking in Africa. The fertility of land in DRC is estimated to be able to support agricultural production, which would be sufficient to feed the whole African population. So what goes wrong with the African countries?

My work experience in West Africa may shed some light on the high price of local food commodities. Rice is the most consumed food commodity as energy source for the population in the region. During my stay in West Africa, with a stomach used to Thai Jasmine rice with a strong aroma, I found the rice I ate in restaurants in West Africa stickier in texture and more appealing than Basmati rice consumed in Europe, which is bland in aroma and less soft in texture. I found out the reason later. During the outbreak of the COVID19 pandemic, I went to supermarkets with my flatmates to replenish our stock of food to get prepared for extreme situations. With a liking for rice I have had for dinner in Côte d'Ivoire, I picked up a bag of local produce but my Ivorian flatmate told me that the imported rice from Vietnam and other regions is better but cheaper. It was at that very moment I realised that I had been savouring rice from Far East rather than local produce for numerous dinners in restaurants and takeaways. Better quality and a lower price tag render imported rice a preference for the local population, and an advantage over locally grown. This poses a serious problem: high as the demand for rice in West Africa, the demand of the locally grown remains low, which reduces the incentive for local farmers to grow, and for rice producers to invest in rice production.

In the first place, it does not make sense that imported rice has an advantage in price compared to the local one. It can be explained at different levels. On the other side of the globe, to ensure the food independency (i.e. sufficient production to feed your own population), a number of countries, however, have been implementing policies to support food production, including subsidies to farmers. As a result, the production can become more than sufficient to feed their own population, but also enables export to other countries at a low price. Therefore, cheap imported food products are easily found on the African market.

Apart from the availability of foreign produce, the cost of production in Africa is high. For some food commodities, there is a high dependency on imported seeds and agricultural inputs, such as fertilisers. The war between Russia and Ukraine has driven up the cost of these imported inputs due to surging shipping cost, exacerbating the burden of farmers in Africa.

Transport cost is another hindering factor. Infrastructure in Africa is far less developed. Unlike developed countries, railway network is far from adequate. It only covers a small fraction of the African continent. Access to agricultural production sites can be difficult as roads are not well built. As a result, the African producers face two dilemmas. Their products cannot reach further markets and can only be sold in their villages and neighbourhood, reducing their incentive to invest in their production for a larger quantity and better quality. Transport cost is relatively high, driving up the price of local products, which fail to compete with foreign products in the local market.

Incapacity to produce enough food for the population could have a serious aftermath on the country. Due to the high dependency on imported food, African countries encounter food crisis when there is a fluctuation in food price on international markets. In 2008, due to a surge of rice price on international markets, the population of West Africa was hit by food insecurity. In Senegal, the rice price doubled in four months, between January and April 2018, while it increased in Mali and Benin by 50%. However, the local production alone could not feed the local population. City dwellers in countries with high rice consumption spent a quarter of their income on rice. The spike in rice price in 2018 had serious implications on food security at household level.[5] When a considerable amount was spent on rice, expenditure on other aspects, such as education and healthcare, was compromised. As can be seen from the crisis in 2008, food independency has a significant impact on people's livelihood in developing countries.

The recent professional experience as a development worker has enabled me to understand that, to foster the development of a country, an integrated approach is necessary. Concerted efforts and actions from

[5] https://www.oecd.org/swac/publications/47853480.pdf

different sectors should be orchestrated to create synergies and complementarities among themselves. Therefore, initiatives contributed by stakeholders from different sectors are more and more designed and implemented. One of the examples is home-grown school feeding programme (HGSF). For school pupils in Hong Kong, school meals are not preferred as their lunch. When I was in the first year of secondary school, my classmates and I were always jealous of the pupils from higher income families, who could go to restaurants and choose their lunch. Unlike Hong Kong, school meals in developing countries is not a matter of choice, but a tool to contribute to local development. School feeding aims to provide nutritious meals to school pupils, preventing them from hunger during school days.

In recent decades, school feeding has been linked to other aspect of local development. Originally ingredients used for school feeding in developing countries were mostly sourced from food donation from donor countries. In recent decades, stakeholders of school feeding advocate for a linkage between local agriculture and school feeding. Such programme is called HGSF. The food demand of school feeding becomes a stable market for local farmers. To become the supplier of school feeding, farmers will have the incentive to invest in their farming for greater and better quality products. The linkage can contribute to the increase in the income of local producers and thus stimulate local economy. In addition, free school meals provide incentives for children to go to school. In Africa, women play an important role in the production of food crops. A certain percentage of quota is set in some countries to procure food for HGSF from women producers, supporting the production of women producers and thus empowering them economically. In addition, parents send their kids to school as they can have the most nutritious meal of the day at school during school days. When I went on a mission in the north of Côte d'Ivoire, I was told by a mother that she did not want to send her kid to school, because she was afraid that her kid could not get any food during classes. She had to keep him with her when working in the farm, so that her little boy could be fed. However, once she came to know that there are free meals in schools, she was willing to send her kid to school. Therefore, the little child was able to receive primary education. Having school meals can also help school pupils to better learn and concentrate in classes.

Apart from education, food security, nutrition and livelihood, school feeding can foster the other aspects in favour of local development, such as gender equality and health. In some countries, families with limited resources prefer sending boys to schools. School meals could be an incentive to convince them of also sending girls to schools, allowing them to reduce food expenditure on their girls. Therefore, school feeding could contribute to gender equality. School feeding can be also combined with health services for school pupils, including deworming. As a whole, HGSF with combined activities supports the development of a country from different aspects. According to WFP, every 1 US dollar invested in school feeding in Ghana yields a return of 3.3 US dollars,[6] demonstrating the effectiveness of school feeding for socio-economic development.

Working as a development worker in developing countries offers me opportunities to have contact with local stakeholders in the field. I feel excited every time when I am on a mission. Field observation and discussion with officials, local communities and local NGO partners provide me with much insights on their perspectives and the needs of beneficiaries. In addition, the hospitality of local communities is far more than what I had expected. Despite their difficult livelihoods, they always prepare their best food to welcome us. During my visit to a village in the north of Côte d'Ivoire in mid-2021, the local community offered us a little sheep. Due to the rules of our office, we cannot receive any gift higher than a certain amount. Moreover, the little sheep looked too cute to be served on our table. Fortunately, the local authority took over our new friend, so that we did not need to think how to deal with it. Moreover, listening to the stories of beneficiaries and their feedback to the project supporting their livelihood is also one of the moments I enjoy the most in the field. They are also my teacher, telling me what works and what does not work in the projects. The knowledge collected from them could assist me in tailoring further support for them in the upcoming projects. The direct interaction with local partners and communities enables me to better understand their needs, which are included during the design of upcoming projects.

[6] World Food Programme (2018). School Feeding in Ghana. Investment case: cost-benefit analysis report: https://ghana.un.org/sites/default/files/2019-10/School%20Feeding%20in%20Ghana_Investment%20Case_Cost%20Benefit%20Analysis%20Report.pdf

Therefore, the projects will be able to better support the local communities to overcome their challenges.

Many people may wonder how we can work in an international organization on humanitarian and development assistance. As there is only little information in Hong Kong on working in international organizations or being engaged in humanitarian and development sector, I am contacted from time to time by young people on Linkedin, who wish to hear my experience and obtain hints on getting a job in the sector. Firstly, many may think that studying international relations or development studies is a must for working in this sector. However, the humanitarian and development stakeholders work on different issues related to 17 sustainable development goals (SDG), which were set up in 2015 by the United Nations General Assembly and are intended to be achieved by 2030. Some of the SDGs are related to poverty reduction, public health, gender equality, agriculture, urban planning, environment and education. Therefore, people with relevant expertise are often sought in the sector, such as economists, nutritionists, agronomists and jurists. In fact, the careers in this sector open the door for a lot of expertise. Secondly, practical experience gained during bachelor or master study, such as through internship, could be useful to pursue your career in this sector. There are a few types of key stakeholders in this sector, who could offer relevant job opportunities:

1. International organzations, such as United Nations agencies
2. Regional Institutions, such as European Union
3. NGOs
4. National Development agencies, which are created by their countries to support the development of their target countries, such as United States Agency for International Development and Deutsche Gesellschaft für Internationale Zusammenarbeit for Germany
5. Development banks, which provide fundings or grants to their partner countries. Examples are the World Bank and the Asian Development Bank
6. Private International Development consulting firms

Doing an internship or volunteer programme can provide you with insights on the specific thematic areas (such as public health, education, human rights) and practical knowledge. And of course, It enables you to understand how the organzations work and how the relevant careers are like. The people you get to know during your internship can share information on relevant job opportunities with you in the future, so that you do not miss any vacancy that may be of your interest. Learning foreign languages, which are spoken in the target countries of the stakeholders, could be a bonus. Communication with national and local stakeholders may require using their languages, such as French, Portuguese, Spanish and Arabic. Speaking these languages fluently may be necessary, especially for those who wish to work in this field. Last but not least, reaching out to people who work in this sector and getting first-hand information on the organization and position in which you are interested. The people working in development sector are very willing to share information and provide advice. I was very lucky to be able to discuss with experienced development workers during my internships. Therefore, I have been doing the same as they do to support young graduates who wish to start their career in this sector, especially those from Hong Kong.

I feel so blessed that, despite a lot of difficulties encountered in the past twenty years, I have been able to have different experiences in the world and meet interesting people who can be considered to have guided me in my study and career development. Working in the development sector is not an easy decision, as there are sacrifices to make (such as staying in a place far from your family and friends). However, the satisfaction from your work and the relevant experience could be unforgettable. If you are the one who wishes to explore the humanitarian and development sector, please be brave and prepare yourself to take up the challenges.

Acknowledgement

I would like to say thank you to my secondary school English teacher, Belinda Wong who provided me with comments and feedback to improve this chapter.

https://doi.org/10.1142/9789811276736_0013

Chapter 13

Chemical Pollution in Agricultural Fields: How it Impacts Our Environment and What We Actually Eat?

Martin Tsz-Ki Tsui

School of Life Sciences, Earth and Environmental Sciences Programme, State Key Laboratory of Agrobiotechnology, Institute of Environment, Energy and Sustainability, The Chinese University of Hong Kong
mtktsui@cuhk.edu.hk

1. Introduction

Human population has a current estimate of about eight billion. This number has been approximated to reach 10 billion by 2055 (Gerland *et al.*, 2014). The ability to feed everyone on this planet will become a major concern. Agriculture has been in human history for a very long time. Various techniques have been developed over different continents and they evolve over time. Agriculture was not modernised until after the industrial revolution by the western world in the 1800s. Since then, different new technologies have been introduced to increase crop and animal production (O'Brien, 1977).

Nowadays, in many parts of the world including some developing countries such as mainland China, many advanced technologies in maximising agricultural productivity have been adopted. However, it has also been widely observed that the high productivity causes various environmental issues (Almaraz *et al.*, 2018).

Current agricultural practice has compromised our environment. Three examples are summarised diagrammatically (Figure 1). First of all, large amount of fertilisers is used in agricultural fields (Tilman *et al.*, 2001) in order to make sure crops grow well. Secondly, pests are widespread in agricultural fields and many pesticides including insecticides and herbicides to suppress the growth of these nuisances are used. These chemical residues can contaminate the soil, surface, groundwater, and crops. They also impact other non-target biota (Silva *et al.*, 2019). Thirdly, adding different chemical components and disturbing the natural lands may lead to indirect consequences such as heavy metal pollution (Rothenberg *et al.*, 2011). Similar to metals, microplastics (defined as

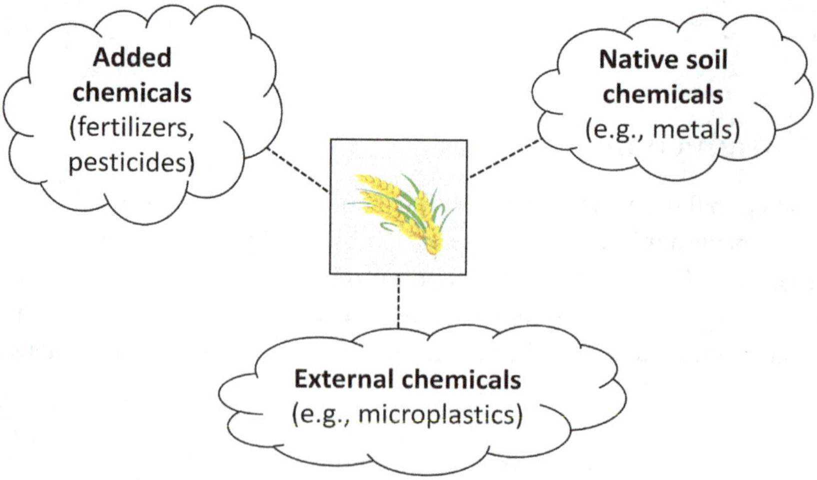

Figure 1. Illustration of the three different environmental chemicals present in common agricultural fields.

plastic debris with a size less than 5 mm) have become problematic for our environment in recent years. They are found all over the environment including agricultural fields (Ullah *et al.*, 2021).

Environmental sustainability is a global issue today. Safeguarding our environment to ensure long-term food security and safety is becoming a key issue to preserve limited resources and human health among many nations. It is the aim of this chapter to provide a brief overview of different types of agricultural pollution and how they impact our environment, food safety, and security.

2. Four Major Agroecosystem Problems

2.1. *Excessive fertilisers*

The invention of fertilisers increased the crop production worldwide but at the same time increased the release of artificially fixed nitrogen and mined phosphorus to the environment. Roughly 20–30% of these nutrients would be taken up by the crop, meaning 70–80% of the added nutrients would end up in the environment, including soil, runoff, groundwater, nearby aquatic systems, and air if the nutrients can be volatile (e.g., ammonia), representing the majority of non-point source pollution in the environment (Carpenter *et al.*, 1998). One of the adverse consequences is the enrichment of the excessive nutrients in the aquatic systems leading to excessive growth of aquatic plants and algae (Daniel *et al.*, 1998). With the decay of these algal biomasses, the water would end up in low levels of dissolved oxygen and can cause massive death of aquatic biota.

Studies have found that mainland China is the largest fertiliser user in the world, followed by other large agricultural nations such as the US (Novotny *et al.*, 2010). Since the majority of applied nutrients may not be actually taken up by the crops, reduction in their uses should be persuaded. Reduced use of fertilisers should benefit the quality of the plants but should not cause any health issue in human consumption.

2.2. *Pesticides*

Pests including unwanted insects and weeds have been the major threats to many agricultural systems worldwide. The invention of various pesticides including insecticides and pesticides spurred their uses. Production of pesticides have been found to increase along with the use of fertilisers in the last century (Tilman *et al.*, 2001). Similar to fertilisers, the majority of pesticides may end up in the environment, some may pose larger concern due to their persistence (up to years or decades) and mobility (e.g., contaminating groundwater). One such notorious example would be the use of agent orange by the US military in Vietnam (Olson and Morton, 2019). Use of agent orange was later proven to cause serious health issues of many Vietnamese people including cancer, birth defects, and severe psychological and neurological problems.

The use of pesticides can cause concerns for human health. First, workers applying them in the field would be exposed to these agrochemicals through various routes such as respiratory and skin contact. Recent court cases in the US of using the herbicide Roundup (glyphosate) raised global concern of long-term exposure to these agrochemicals. Second, the drift of pesticides to the environment may cause toxic effects to "non-target" organisms. One such attribution is to the global decline of honeybees and amphibians (Relyea, 2005). While sometimes the active ingredient itself is not very toxic to animal consumers, the additive chemicals (e.g., surfactant to glyphosate formulation, Roundup) have been found to be more potent to animal organisms (Tsui and Chu, 2003). Third, the residue of pesticides on the crop would lead to human's indirect uptake of toxic pesticide residues. Many of these pesticides are organic in nature. They may not be easily rinsed off by using just tap water. Thus, farms using less or no pesticides are widely popular. One such business model would be organic farming.

2.3. *Toxic metals*

Metals are naturally existing elements and may be inherently present in many agricultural soils, with some areas higher in concentrations due to geological deposits. Human activities may enhance the bioavailability of these toxic elements and some of them can be taken up at high levels by the crops in the field. Metals can also possess organismal inhibitory capability, as such they can be used as insecticides, herbicides, and fungicides. Depending on the extent of the uses, their concentration can be highly elevated in the agricultural soils and can be widely taken up by the crops.

In the last one to two decades, a lot of research has shown that humans can take up toxic methylmercury by consuming rice, in addition to fish, posing a risk to inland populations in which they do not consume fish and can have elevated methylmercury levels (Rothenberg *et al.*, 2011). The main reason behind this process is that the rice fields are inundated with water for growing rice and this would create an anaerobic condition in which certain anaerobic bacteria such as sulfate-reducing and iron-reducing bacteria can convert inorganic mercury to highly toxic organic methylmercury in the soils of rice field. Subsequently, methylmercury can be taken up by the rice plant through the root systems. One of the strategies on reducing methylmercury uptake by rice plants would be to use soil amendments such as biochars and activated carbon. These soil amendments have been shown to reduce the anaerobic mercury methylation in inundated soils and thus reduce its uptake by rice plants (Shu *et al.*, 2016).

2.4. *Microplastics*

Microplastics have become one of the hottest environmental topics nowadays due to the massive production and widespread use in human society, regardless of wealth and development status of the country. Recent research has shown that microplastics, being less than 5 mm, are everywhere in our planet including air, soil, water, food, and organisms (Cox *et al.*, 2019; Zhang *et al.*, 2020). In the environment, microplastics have been found in some extreme and remote areas such as the Arctic and deep ocean, illustrating its wide presence.

While early research on microplastics focused on freshwater and marine ecosystems, recent work has switched the focus to terrestrial lands including agricultural fields. One of the most important questions is whether microplastics can find a way to enter the food crops (Ullah *et al.*, 2021) and this is illustrated in Figure 2. So far, little evidence has indicated their potential uptake. For example, Li *et al.* (2020) showed in a

Figure 2. Incorporation of microplastics into our food systems and entering humans.

laboratory experiment that microplastics can enter the crop plants by a "crack-entry mode" pathway in the roots. Other studies have tested the effects of the microplastics on growth performance of plants (Rillig *et al.*, 2019). While these studies found that microplastics could have negative effects on the soil structures and plant performance, their manipulated levels of microplastics (e.g., 2% dry mass in soil) in the control experiments are considered to be not environmentally realistic (de Souza Machado *et al.*, 2019). More research is needed in this area for detection of natural microplastics in the food items to make a more definitive conclusion.

Summary

Due to years of human agriculture activities along with direct and indirect releases to the environment, many natural and manmade chemicals are already widely present in the environment, including remote and extreme ecosystems investigated. Therefore, it is not surprising that agricultural fields, which are considered as "man-made" ecosystems, are fully contaminated with all these chemicals. However, one should be cautious about how these chemicals influence the agricultural lands and whether they accumulate in the crop plants. All natural ecosystems and plants have certain capability in tolerating these chemicals and not all chemicals are highly bioavailable to be taken up by the crops. Careful investigation is still needed to examine the risk and hazard of chemicals to the agricultural fields and plants, that in turn affects our food safety and security.

References

Almaraz, M., Bai, E., Wang, C., Trousdell, J., Conley, S., Faloona, I., and Houlton, B.Z., 2018. Agriculture is a major source of NO_x pollution in California. *Sci. Adv.* 4: eaao3477.

Carpenter, S.R., Caraco, N.F, Correll, D.L., Howarth, R.W., Sharpley, A.N., and Smith, V.H., 1998. Nonpoint pollution of surface waters with phosphorus and nitrogen. *Ecol. Appl.*, 8: 559–568.

Cox, K.D., Covernton, G.A., Davies, H.L., Dower, J.F., Juanes, F., and Dudas, S.E., 2019. Human consumption of microplastics. *Environ. Sci. Technol.*, 53: 7068–7074.

Daniel, T.C., Sharpley, A.N., and Lemunyon, J.L., 1998. Agricultural phosphorus and eutrophication: A symposium overview. *J. Environ. Qual.*, 27: 251–257.

de Souza Machado, A.A., Lau, C.W., Kloas, W., Bergmann, J., Bachelier, J.B., Faltin, E., Becker, R., Görlich, A.S., and Rillig, M.C., 2019. Microplastics can change soil properties and affect plant performance. *Environ. Sci. Technol.*, 53: 6044–6052.

Gerland, P., Raftery, A.E., Ševcíková, H., Li, N., Gu, D., Spoorenberg, T., Alkema, L., Fosdick, B.K., Chunn, J., Lalic, N., Bay, G., Buettner, T., Heilig, G.K., and Wilmoth, J., 2014. World population stabilization unlikely this century. *Science*, 346: 234–237.

Li, L., Luo, Y., Li, R., Zhou, Q., Peijnenburg, W.J.G.M., Yin, N., Yang, J., Tu, C., and Zhang, Y., 2020. Effective uptake of submicrometre plastics by crop plants via a crack-entry mode. *Nat. Sustain.*, 3: 929–937.

Novotny, V., Wang, X., Englande Jr., A.J., Bedoya, D., Promakasikorn, L., and Tirado, R., 2010. Comparative assessment of pollution by the use of industrial agricultural fertilizers in four rapidly developing Asian countries. *Environ. Develop. Sustain.*, 12: 491–509.

O'Brien, P.K., 1977. Agriculture and the industrial revolution. *The Economic History Review The Economic History Review New Series*, 30: 166–181.

Olson, K.R. and Morton, L.W., 2019. Long-term fate of agent orange and dioxin TCDD contaminated soils and sediments in Vietnam hotspots. *Open J. Soil Sci.* 9: 1–34.

Relyea, R.A., 2005. The lethal impact of Roundup on aquatic and terrestrial amphibians. *Ecol. Appl.*, 15: 1118–1124.

Rillig, M.C., Lehmann, A., de Souza Machado, A.A., and Yang, G., 2019. Microplastic effects on plants. *New Phytol.*, 223: 1066–1070.

Rothenberg, S.E., Feng, X., Dong, B., Shang, L., Yin, R., and Yuan, X., 2011. Characterization of mercury species in brown and white rice (*Oryza sativa* L.) grown in water-saving paddies. *Environ. Pollut.*, 159: 1283–1289.

Shu, R., Wang, Y. and Zhong, H., 2016. Biochar amendment reduced methylmercury accumulation in rice plants. *J. Hazard. Mater.* 313: 1–8.

Silva, V., Mol, H.G.J., Zomer, P., Tienstra, M., Ritsema, C.J., and Geissen, V., 2019. Pesticide residues in European agricultural soils — A hidden reality unfolded. *Sci. Total Environ.*, 653: 1532–1545.

Tilman, D., Fargione, J., Wolff B, D'Antonio, C., Dobson, A., Howarth, R., Schindler, D., Schlesinger, W.H., Simberloff, D., and Swackhamer, D., 2001. Forecasting agriculturally driven global environmental change. *Science*, 292: 281–284.

Tsui, M.T.K. and Chu, L.M., 2003. Aquatic toxicity of glyphosate-based formulations: comparison between different organisms and the effects of environmental factors. *Chemosphere*, 52: 1189–1197.

Ullah, R., Tsui, M.T.K., Chen, H., Chow, A.T., Williams, C., and Ligaba-Osena, A., 2021. Microplastics interaction with terrestrial plants and their impacts on agriculture. *J. Environ. Qual.*, 50: 1024–1041.

Zhang, Y., Kang, S., Allen, S., Allen, D., Gao, T., and Sillanpää, M., 2020. Atmospheric microplastics: A review on current status and perspectives. *EarthSci. Rev.*, 203: 103118.

https://doi.org/10.1142/9789811276736_bmatter

Index

Printed in the USA
CPSIA information can be obtained
at www.ICGtesting.com
JSHW010754180524
63072JS00002B/6

9 789811 276729